Automated Vehicles Symposium 2016

Drivers. Vehicles. Infrastructure.

San Francisco, California, USA
21 July 2016

ISBN: 978-1-5108-5183-2

Printed from e-media with permission by:

Curran Associates, Inc.
57 Morehouse Lane
Red Hook, NY 12571

Some format issues inherent in the e-media version may also appear in this print version.

Printed by Curran Associates, Inc. (2018)

For permission requests, please contact Association for Unmanned Vehicle Systems International at the address below.

Association for Unmanned Vehicle Systems International
2700 South Quincy Street
Suite 400
Arlington, VA 22206
USA

Phone: (703) 845-9671
Fax: (703) 845-9679

www.auvsi.org

Additional copies of this publication are available from:

Curran Associates, Inc.
57 Morehouse Lane
Red Hook, NY 12571 USA
Phone: 845-758-0400
Fax: 845-758-2633
Email: curran@proceedings.com
Web: www.proceedings.com

Contents

POSTER SESSIONS

APPENDIXES

TRANSPORTATION RESEARCH BOARD
2017 EXECUTIVE COMMITTEE OFFICERS

Chair: Malcolm Dougherty, Director, California Department of Transportation, Sacramento

Vice Chair: Katherine F. Turnbull, Executive Associate Director and Research Scientist, Texas A&M Transportation Institute, College Station

Division Chair for NRC Oversight: Susan Hanson, Distinguished University Professor Emerita, School of Geography, Clark University, Worcester, Massachusetts

Executive Director: Neil J. Pedersen, Transportation Research Board

TRANSPORTATION RESEARCH BOARD
2017–2018 TECHNICAL ACTIVITIES COUNCIL

Chair: Hyun-A C. Park, President, Spy Pond Partners, LLC, Arlington, Massachusetts

Technical Activities Director: Ann M. Brach, Transportation Research Board

David Ballard, Senior Economist Gellman Research Associates, Inc., Jenkintown, Pennsylvania, *Aviation Group Chair*

Coco Briseno, Deputy Director, Planning and Modal Programs, California Department of Transportation, Sacramento, *State DOT Representative*

Anne Goodchild, Associate Professor, University of Washington, Seattle, *Freight Systems Group Chair*

George Grimes, CEO Advisor, Patriot Rail Company, Denver, Colorado, *Rail Group Chair*

David Harkey, Director, Highway Safety Research Center, University of North Carolina, Chapel Hill, *Safety and Systems Users Group Chair*

Dennis Hinebaugh, Director, National Bus Rapid Transit Institute, University of South Florida Center for Urban Transportation Research, Tampa, *Public Transportation Group Chair*

Bevan Kirley, Research Associate, Highway Safety Research Center, University of North Carolina, Chapel Hill, *Young Members Council Chair*

D. Stephen Lane, Associate Principal Research Scientist, Virginia Center for Transportation Innovation and Research, *Design and Construction Group Chair*

Ram M. Pendyala, Frederick R. Dickerson Chair and Professor of Transportation, Georgia Institute of Technology, *Planning and Environment Group Chair*

Joseph Schofer, Professor and Associate Dean of Engineering, McCormick School of Engineering, Northwestern University, Evanston, Illinois, *Policy and Organization Group Chair*

Robert Shea, Senior Deputy Chief Counsel, Pennsylvania Department of Transportation, *Legal Resources Group Chair*

Eric Shen, Director, Southern California Gateway Office, Maritime Administration, Long Beach, California, *Marine Group Chair*

William Varnedoe, Partner, The Kercher Group, Raleigh, North Carolina, *Operations and Preservation Group Chair*

TRANSPORTATION RESEARCH CIRCULAR E-C222

Automated Vehicles Symposium 2016

Summary of a Symposium

July 21, 2016
Hilton San Francisco Union Square
San Francisco, California

Cosponsored by
Transportation Research Board of the National Academies of Sciences, Engineering, and Medicine
Association for Unmanned Vehicle Systems International

Proceedings Supported by
Federal Highway Administration

Rapporteur
Katherine F. Turnbull
Texas A&M Transportation Institute

July 2017

Transportation Research Board
500 Fifth Street, NW
Washington, D.C.
www.TRB.org

TRANSPORTATION RESEARCH CIRCULAR E-C222
ISSN 0097-8515

The **Transportation Research Board** is one of seven programs of the National Academies of Sciences, Engineering, and Medicine. The mission of the Transportation Research Board is to provide leadership in transportation innovation and progress through research and information exchange, conducted within a setting that is objective, interdisciplinary, and multimodal.

The **Transportation Research Board** is distributing this E-Circular to make the information contained herein available for use by individual practitioners in state and local transportation agencies, researchers in academic institutions, and other members of the transportation research community. The information in this circular was taken directly from the submission of the authors. This document is not a report of the National Academies of Sciences, Engineering, and Medicine.

Transportation Research Board
500 Fifth Street, NW
Washington, D.C.
www.TRB.org

Introduction

Katherine F. Turnbull
Texas A&M Transportation Institute

SYMPOSIUM OVERVIEW

The 2016 Automated Vehicle Symposium (AVS) was held in San Francisco, California, on July 19–21, 2016. The symposium was organized and conducted through a partnership between the Transportation Research Board (TRB) of the National Academies of Sciences, Engineering, and Medicine and the Association for Unmanned Vehicle Systems International (AUVSI). The symposium featured keynote speakers, presentations on current projects and programs, 22 concurrent breakout sessions, and ancillary meetings.

The development, testing, building, and deployment of autonomous vehicles continue at a rapid pace. These technologies will have a major impact on transportation safety and mobility, as well as the environment and urban forms. The introduction of these technologies also has the potential to disrupt the transportation system.

The 2016 AVS provided the opportunity for communication, collaboration, and information sharing on a wide range of topics, including potential public policy, safety and security, ethics, and equity concerns, as well as technology innovations and applications. The 2016 AVS attracted almost 1,200 participants from the United States and other countries. The growth in attendance from 2013 to 2016 reflects the increased interest in connected and automated vehicles (CAVs) and autonomous vehicles. Approximately 250 people attended the 2013 symposium. Attendance increased to 550 people in 2014, 862 people in 2015, and almost 1,200 people in 2016.

The 2016 symposium attracted a global audience, with attendees from 25 countries. Approximately 21% of attendees came from outside the United States. Japan had the largest number of attendees, with 60 people. The United Kingdom, South Korea, Germany, and Canada followed, with 21 to 23 attendees each. New countries represented at the 2016 symposium included Mexico, Poland, Russia, and Singapore.

The 2016 symposium attracted participants from 40 states, an increase of one state from 2015. The actual states changed, however, with some dropping out and some joining. The state with the largest number of attendees has changed with the location of the symposium. In 2014, California hosted the symposium and recorded the most participants. With the change in venue to Ann Arbor in 2015, Michigan led all states in attendees. With the return to San Francisco in 2016, California regained the top spot. Washington, D.C., and Virginia ranked third and fourth, respectively, for all 3 years.

Attendees at the 2016 AVS represented a wide range of public agencies and private-sector groups. Individuals from federal, state, regional, metropolitan, and local agencies participated in the symposium, as did representatives from technology and software companies, automotive original equipment manufacturers (OEMs), and tier 1 suppliers, consulting firms, and other industries. Researchers and faculty from universities and research institutes, as well as from nonprofit organizations and other groups, also participated in the symposium.

REPORT ORGANIZATION

This report presents the proceedings from the 2016 AVS. The report follows the general symposium agenda. The presentations by speakers in the general sessions are summarized including the keynote addresses and the highlights from the 22 breakout sessions. A list of the posters presented in two sessions is provided. The appendixes provide a description of the key topics covered in the breakout sessions.

Some of the symposium speakers provided their PowerPoint presentations or other supporting materials, which are available at http://www.automatedvehiclessymposium.org/home.

EDITOR'S NOTE

The organizers of the Automated Vehicles Symposium 2016 endorse the term "automated" in reference to vehicles that have advanced driving capabilities, rather than the term autonomous. However, the terminology in these proceedings remain faithful to the speakers' own words, and so retain the terms "self-driving," "autonomous," and "automated," according to the speakers' or authors' original choice.

PUBLISHER'S NOTE

The preparation of this report was supported by the Federal Highway Administration (FHWA) and TRB. The views expressed in this summary are those of the individual speakers and discussants, as attributed to them, and do not necessarily represent the consensus views of the symposium participants, the volunteer organizers, TRB, or the National Academies of Sciences, Engineering, and Medicine. This e-circular has not been subjected to the formal TRB peer-review process.

Opening Session

JANE LAPPIN
Transportation Research Board Intelligent Transportation Systems Committee
Toyota Research Institute, presiding

WELCOME FROM THE TRANSPORTATION RESEARCH BOARD

Jane Lappin, *Chair, Toyota Research Institute*

Jane Lappin provided a welcome on behalf of TRB of the National Academies of Sciences, Engineering, and Medicine. She summarized the background to the 2016 AVS and highlighted elements of the program. Lappin covered the following topics in her presentation:

- Lappin noted that the first symposium was held 5 years ago with 125 participants. Approximately 1,200 people are attending the 2016 symposium. She commented that there have been major research, development, testing, pilots, and demonstrations of automated vehicles (AVs) over the same 5-year period.
- Lappin suggested that while AVs will provide numerous benefits, there are also complex challenges associated with their deployment. The AVS brings together diverse groups from industry, academia, public agencies, and other stakeholders to share information, to discuss issues and opportunities, and to identify future research needs, collaboration strategies, and emerging topics of interest.
- Lappin thanked TRB and AUVSI for cosponsoring the symposium. She recognized and thanked the members of the 2016 symposium executive committee and the almost 200 volunteers who organized the 22 breakout sessions. She noted that TRB is a private nonprofit organization chartered by Congress to provide innovative, research-based solutions to improve transportation. Over 7,000 volunteers participate in TRB's 220 standing committees and task forces, and an estimated 13,000 people attend the Annual Meeting of the Transportation Research Board held in Washington, D.C., each January. TRB also holds conferences and workshops, and publishes numerous reports on key topics to advance transportation.
- Lappin noted that TRB and AUVSI have partnered on the AVS to provide one international event for learning, networking, and developing partnerships on AVs. She thanked the 27 benefactors and exhibitors for their support of the symposium. She also acknowledged and thanked the speakers, the AUVSI staff, the breakout session organizers, and all the participants.

WELCOME FROM THE ASSOCIATION FOR UNMANNED VEHICLE SYSTEMS INTERNATIONAL

Brian Wynne, *Association for Unmanned Vehicle Systems International*

Brian Wynne provided a welcome from AUVSI. He summarized the role of AUVSI and the interest in AVs. He also thanked the various groups and individuals who organized the 2016 symposium. Wynne covered the following topics in his presentation:

- Wynne reported that AUVSI is the world's largest nonprofit organization devoted exclusively to advancing the unmanned systems and robotics community. He noted that AUVSI was pleased to partner with TRB on the 2016 AVS. He thanked TRB, the 2016 executive committee, the 27 benefactors and exhibitors, the volunteers organizing the breakout sessions, the speakers, and the participants.
- Wynne noted that the AVS continues to grow, with more companies, policy makers, educators, and researchers participating. AVs are receiving unprecedented press in the United States and worldwide. He stressed that the AVS is an international event.

WELCOME TO CALIFORNIA

Brian Kelley, *Secretary, California State Transportation Agency*

Brian Kelley provided a welcome from the California State Transportation Agency. He thanked TRB, AUVSI, and all of the volunteers for organizing the symposium. He highlighted activities in California associated with connected, automated, and autonomous vehicles. Kelley covered the following topics in his presentation:

- Kelley noted that the symposium provided an international forum for the discussion of autonomous vehicles. He commented that it is a very exciting time with the rapid advancements in technology impacting all aspects of transportation. People are buying vehicles on the Internet; rideshare companies are providing on-demand services; and transit and transportation agencies are providing real-time information on the status of buses, trains, and traffic conditions.
- Kelley noted that autonomous vehicles may achieve three policy goals. These goals are improving roadway safety, mitigating environmental impacts, and expanding mobility options. He suggested that autonomous vehicles can help achieve the Vision Zero goal of no roadway fatalities, assist in meeting the last-mile needs of transit operators, and provide mobility to individuals who cannot or choose not to drive.
- Kelley noted that California, especially companies in the Silicon Valley, is playing a key role in the development of AVs. Further, the state of California has sponsored research on vehicle automation since the late 1980s at the University of California and other universities in the state. In 2014, the California Department of Motor Vehicles issued regulations for the testing of autonomous vehicles in the state. He noted that 14 companies are currently approved to test autonomous vehicles in the state. He further noted that this ability to test vehicles in a real-world

environment was an important step in the deployment process, but that testing must be conducted in a safe manner.

KEYNOTE ADDRESS

Anthony R. Foxx, *Secretary, U.S. Department of Transportation*

Anthony Foxx provided an update on AV activities underway at the United States Department of Transportation (DOT). He covered the following topics in his presentation:

- Foxx noted the importance of the intersection of technology, innovation, and transportation. He suggested that a dramatic transformation was occurring throughout the world in the way people move. He commented that autonomous vehicles are coming. The choice for governmental agencies is to either act or react. Foxx suggested that acting and preparing the transportation ecosystem to accept these new types of vehicles was critical. He noted that the goal was to be ready to integrate autonomous vehicles safely into the transportation system.
- Foxx identified keys to meeting the goal for safely integrating of autonomous vehicles into the transportation system. He noted that the active involvement of government, industry, and consumers was one key. He stressed the importance of addressing safety concerns, with autonomous vehicles helping to reduce or eliminate the 80% of crashes that are caused by human error. One challenge he cited was to ensure that crashes caused by technology errors or malfunctions does not increase. He stated that the prospect of reducing crashes is of high interest to the U.S. DOT.
- Foxx noted the need for clear lines of responsibility between government, industry, and consumers. He suggested that managing the introduction of autonomous vehicle technology through premarket testing, which would require government and industry to develop requirements, represented a logical approach, as it would allow consumers to know that vehicles on the market had been tested.
- Foxx discussed the need to better define the roles of the federal government and of state governments in testing and deploying autonomous vehicles. He noted that the U.S. DOT received input that having 50 different sets of regulations governing testing, deploying, and operating autonomous vehicles was not logical. He suggested the need to provide more predictability in the marketplace, while still examining many issues in more detail. Foxx reported that the U.S. DOT was working with the American Association of Motor Vehicle Administrators on model state policies. The goal is to ensure that states adopt policies that align with policies in other states and at the federal level.
- Foxx suggested the need for new tools, more nimble approaches, and more flexibility in developing policies and guidance. He said that a new federal advisory committee would be formed later in the year to assist with examining possible issues and approaches, and developing guidelines and policies. It is envisioned that members of the committee will represent diverse backgrounds, expertise, perspectives, and public and private groups.
- Foxx noted that it was an exciting time to be in transportation. He cautioned to not focus just on the excitement of technology, however. He suggested that maintaining a safe and efficient transportation system for people and goods is still the key goal.

GENERAL SESSION

Vehicle Automation Technology Deployment

Bob Denaro
ITS Consultant and Chair, TRB Joint Subcommittee on Challenges and Opportunities for Road Transport Automation, presider

SOCIALLY ACCEPTABLE AI-BASED URBAN DRIVING

Maarten Sierhuis, *Nissan Research Center Silicon Valley*

Maarten Sierhuis discussed current activities at the Nissan Research Center Silicon Valley. He described the complex situations autonomous vehicles will need to navigate and explained the concept of socially acceptable autonomous driving. Sierhuis covered the following topics in his presentation:

- Sierhuis presented videos to highlight some of the difficulties encountered in developing autonomous vehicles. He described how the interaction complexity increases with different types of roads, different environments, and different road users. He noted that autonomous vehicles must be able to operate in all these different situations. He noted that Nissan is conducting tests at the National Aeronautics and Space Administration (NASA) Ames Research Center in Silicon Valley. The Ames facility allows Nissan to test vehicles operating in a closed environment before testing on public roads. One video illustrated autonomous vehicles operating at an intersection.
- Sierhuis described some of the challenges in developing an autonomous vehicle that can operate in complex situations. He noted that autonomous vehicles must obey traffic laws, understand four-way stops at intersections, negotiate right-turn-on-red situations, and recognize other road users. He also noted that autonomous vehicles must understand socially acceptable driving behaviors and must be able to communicate with other road users. Other challenges for autonomous vehicles cited by Sierhuis included navigating road construction sites, work zones, and alternative routes. People and changing environments represented other challenges he illustrated.
- Sierhuis described the concept of socially acceptable autonomous driving. He suggested that driving is a social interaction, which an autonomous vehicle needs to understand. He highlighted these points using videos showing the interaction of vehicles and pedestrians in different situations, noting the importance of body language in understanding the behavior of pedestrians. The first video was from San Francisco. He also presented videos from Tehran, Iran, illustrating different pedestrian behaviors and the presence of numerous motorcycles. Next, he presented a video of the bicycle and automobile traffic in Amsterdam. Sierhuis noted that autonomous vehicles must be able to operate in all these situations and environments.
- Sierhuis stressed the importance of research examining the autonomous systems that communicate a vehicles' intent to other road users. He described research at Nissan focusing on the development of a vehicle intention indicator, which was demonstrated at the Tokyo Motor Show. He presented a video of the concept in operation on a vehicle.

AUTOMATED, CONNECTED, AND ELECTRIC VEHICLES

Jan Becker, *Faraday Future*

Jan Becker described the development of intelligent electric vehicles at Faraday Future. He described the modular platform approach and the prototype vehicle. Becker covered the following topics in his presentation:

- Becker presented a video highlighting the concept vehicle developed by Faraday Future. The company was established in 2014 and is developing manufacturing facilities in Las Vegas, Nevada, and in the San Francisco area. He noted that a variable platform architecture was developed that can be used across all types of vehicles. He described the battery and wheelbase configuration and the motor and powertrain configuration. The modular platform approach provides flexibility for use with different sizes of vehicles.
- Becker described the development of the first prototype vehicle, which is being used as the basis for additional vehicles. He suggested that it was important to ensure that autonomous vehicles provide an enjoyable user experience.
- Becker also stressed the importance of safety and the safe operations of autonomous vehicles. He described the redundancy and monitoring features in the prototype vehicle. He described some of the human factors challenges encountered with the different levels of automation. He also described the advancements in sensors and other technologies that enable the development of autonomous vehicles. Becker noted that Faraday Future has been approved for testing in California.

BRINGING AUTONOMOUS VEHICLES INTO PRODUCTION: AN AUTOMOTIVE ORIGINAL EQUIPMENT MANUFACTURER PERSPECTIVE

Colm Boran, *Ford Motor Company*

Colm Boran described the Ford approach to developing autonomous vehicles. He discussed the Ford product development process and the importance of collaboration. Boran covered the following topics in his presentation:

- Boran reviewed the Society of Automotive Engineers (SAE) automation levels, noting that a driver is not required at Level 4 when automation is engaged, while Level 5 is full automation. He described the Ford Level 4 autonomous vehicle configuration. The two major subsystems are the virtual driver subsystem and the autonomous vehicle platform. The virtual driver subsystem provides the sensing functions using cameras, radar, light detection and ranging (lidar), Global Positioning Systems (GPS), and other technologies. The autonomous vehicle platform provides motion control using hybrid electric propulsion, brake control, and steering control. Level 4 requires redundancy in braking and steering control, sensors, and power. The configuration platform is intended for use with all Ford vehicle makes and models.
- Boran noted that it is anticipated that autonomous vehicles will experience normal driving situations and unusual driving situations. Autonomous vehicles must be able to function in all settings. In anticipation of these needs, Ford has developed state-of-the-art sensing

algorithms. He presented a video that highlighted the sensing capabilities of a research vehicle, including the use of lidar.

- Boran noted that there is a big difference between building one autonomous vehicle and building millions of high-quality autonomous vehicles that meet the needs of diverse customers. Vehicles are complex, as in the driving environment. He commented that automobile manufacturers have a long history of building vehicles, knowing the environments, and understanding the technology. He suggested that with test tracks, development facilities, and assembly plants, the automobile industry is well positioned to produce autonomous vehicles.
- Boran described Ford's project development process, which includes cataloging, anticipating, and testing and validating. He noted the decades of experience cataloging vehicle design and operating data. He reported that Ford autonomous vehicles are being designed to drive in the same environment as today's vehicles. Attributes and noise factors include vehicle performance, dynamics, ride, as well as temperature and other weather conditions. This information is used to anticipate changes and to develop new scenarios. Testing and validating these scenarios includes computer simulation, closed-loop simulation, test track operations, and real-world driving. Boran discussed the use of and features of the MCity test facility in Michigan for testing autonomous vehicles.
- Boran discussed the importance of collaboration in developing connected, automated, and autonomous vehicles. He highlighted examples of activities where collaboration is key, including the development of functional automation level descriptions, safety principles for each automation level, functional safety ratings, and objective test methods. He further noted that coordinating and collaborating with regulatory bodies and other agencies and groups was also important. He stressed the importance of developing vehicles that achieve high levels of safety for a wide range of tested conditions throughout the world.

CONNECTED–AUTOMATED TRUCKS

Josh Switkes, *Peloton Technology, Inc.*

Josh Switkes discussed truck platooning. He described the importance of trucking to the economy, the evolution of manual steering truck platooning to automated trucks, and elements of safe deployment. Switkes covered the following topics in his presentation:

- Switkes noted that the trucking network powers the U.S. economy. Trucking is a $700-billion connected industry that carries approximately 70% of freight in the country. He also noted that trucking has a low margin, with concentrated expenses in fuel, crashes, and labor. For example, on an annual basis, the trucking industry spends close to $100 billion on diesel fuel, $60 billion on costs associated with crashes, and $300 billion on labor. According to Switkes, automation can assist in reducing all of these costs.
- Switkes provided a perspective on the right increments of automation for trucks. The first step is manual steering platooning. The benefits from this step include crash prevention and fuel savings. Field data collected on manual steering platooning will assist in moving to the next increments, which is fully automated following platooning. This step provides labor cost reductions and additional fuel savings. The standalone Level 3–4 automation will provide further labor reductions and will allow for the optimization of all operations.

- Switkes presented a video illustrating truck platooning in operation. He suggested that platooning can drive the development and the safe deployment of higher levels of automation. Key elements of the connective trucks include a vehicle-to-vehicle (V2V) wireless link, radar on both trucks, and a wireless link to the network operations center. The active braking provides nearly synchronized braking. With the platooning, both drivers steer and both trucks save fuel. The network operations center (NOC) and real-time cloud supervision allows platooning only when it is safe, where it is safe, and how it is safe. It also provides dynamic adjustments to conditions. He noted that trucks are both producers and consumers of data.
- Switkes discussed the NOC elements for safe deployment of truck platooning. The first element cited was providing approval based on the right roads, the known roads, and the right conditions. Describing truck platooning on limited access roadways that have been used before in safe weather and operating conditions was noted. The second element was making adjustments as needed, including modifying following distance and other parameters. The third element was coordination in finding and maintaining platooning partners. The final element was data collection for the development of more advanced systems.
- Switkes highlighted the major roadway network used by trucks in the United States, which represents approximately 4,500 mi of roads. He noted that the focus of truck platooning is anticipated to be long stretches of the Interstate system, not in urban areas.
- Switkes described moving from truck platooning to truck automation. He noted that in the platooning phase, Peloton has focused on developing the technology, the team, the brand, and the partners. It has also engaged regulators and customers. Data-driven development and deployment toward automation include operational intelligences, real-time supervision, up-to-date road data, and simulation and machine learning sources. The ongoing updating of road data is a key element in all these activities. Products from these elements include local sensing and decisions, planning and supervision, lateral control, and longitudinal control. Future steps include automated following-vehicles and standalone automation.
- Switkes reviewed the states that have approvals for trials and demonstrations or have expressed interest in Level 1 manual steering truck platooning. States with approval for trials and demonstrations include Alabama, California, Florida, Michigan, Nevada, Ohio, Texas, Utah, and Virginia.

GENERAL SESSION

Nontechnical Issues

STEVEN SHLADOVER
University of California PATH Program and Chair,
TRB Vehicle–Highway Automation Committee, presider

ETHICS OF AUTONOMOUS VEHICLES: BEYOND WEIRD CRASH DILEMMAS

Patrick Lin, *California Polytechnic State University*

Patrick Lin provided an introduction to ethics associated with autonomous vehicles beyond weird crash dilemmas. He discussed possible ethical issues related to testing, licensing, and insuring autonomous vehicles, as well as social and legal concerns. Lin covered the following topics in his presentation:

- Lin discussed the importance of technology ethics. He noted that while examining the benefits of autonomous vehicles is important, it is only part of the story. He cited the example of cancer fighting drugs, which may have harmful side effects for some people. He suggested that ethics sometimes require society to make difficult value judgements, such as jobs versus environmental impacts and speed versus safety.
- Lin presented one possible dilemma that an autonomous vehicle might encounter in having to swerve to either the right or the left. Swerving to the left the autonomous vehicle would hit a motorcyclist not wearing a helmet and swerving to the right the vehicle would hit a motorcyclist wearing a helmet. He suggested that it was useful to discuss these types of potential situations and the ethics of potential decisions. He commented that science experiments do not always replicate the real world, but that they are still useful.
- Lin noted that people make decisions every day that put others at risk. For example, a driver giving extra room to a school bus with children may come closer to bicyclists and pedestrians. He suggested that drivers make these types of decisions on a regular basis and that autonomous vehicles may have to make similar decisions. Lin suggested that ignoring ethics could have negative consequences on public perceptions, technology adoption and market share, lawsuits, and regulatory action.
- Lin discussed seven ethical considerations with the development and deployment of autonomous vehicles. The first consideration related to testing and if it is ethical to beta-test autonomous driving on public roadways. He suggested that these tests could be considered human subject research, which requires the development of research protocols and approval by an institutional review board (IRB). He further noted that other road users, including bicyclists and pedestrians, would be impacted by these tests.
- Licensing for autonomous vehicles was the second area discussed by Lin. He asked if it was enough for an autonomous vehicle to pass the same test that human drivers take or if there should be other requirements?
- Navigation was the third ethics topic described by Lin. He suggested that examining the potential large-scale effects of autonomous navigation was important. One question might be

who is liable if a car takes you into a forbidden travel zone? He commented that issues are emerging with navigation systems routing vehicles through residential areas to bypass congestion on major roadways. Residential roads were not designed to accommodate high traffic levels, putting residents at risk.

- A fourth ethical consideration discussed by Lin was services. He questioned if autonomous vehicles would keep driving past advertisements or specific commercial establishments who had paid for this service to attract customers.
- Insurance was the fifth area discussed by Lin. Possible ethical issues include how to account for new risks and unpredictability that emerge from the complexity of autonomous vehicles and possible new modes of insuring micro-risks. He suggested that the insurance industry will likely have to evolve and innovate with the advent of autonomous vehicles.
- Lin described some of the potential social ethical issues associated with autonomous vehicles. He noted that traffic fines help fund police, fire, education, and social services in many communities. If traffic fines disappear because autonomous vehicles do not speed or park illegally, funding for these services will also disappear and will need to be replaced by other sources. Additionally, if fatalities from vehicle crashes decline, so will organ donations. He stressed that while none of these factors should dissuade moving forward with technologies, they are disruptions that need to be considered.
- Ethics related to legal issues was the final area Lin discussed. He noted that liability and privacy were two major concerns with autonomous vehicles. Determining who is liable in the case of a crash is one issue. The other issue concerns programming a vehicle to protect its occupants. If a vehicle is programmed to protect its occupants, could it be considered premediated if a fatality occurred as the result of a crash where the vehicle is trying to avoid its occupants being harmed?
- Lin noted that Breakout Session 10 addresses ethics. He described the four session panels focusing on the importance of ethics, values and weights, licensing and testing, and consumer perceptions.
- Lin commented that autonomous vehicles are more than just the technology. He suggested that automobiles are a very iconic part of the American landscape and can be considered a sociopolitical system with all the laws, regulations, and customs governing their use. Disruptive technology also disrupts this system. He suggested that disruption was not bad, but that it was important to manage the change and to continue to discuss these topics.

ARE CONSUMERS READY AND WAITING FOR AUTOMATED VEHICLES?

Kristin Kolodge, *J.D. Power and Associates*

Kristin Kolodge discussed recent J.D. Power market research on vehicle technologies, including advanced driver assistance systems (ADAS). She highlighted consumer interest in different ADAS features and described some of the risks associated with consumer adoption resulting from variations in the industry. Kolodge covered the following topics in her presentation.

- Kolodge described both the increasing penetration of ADAS in new vehicles and the increasing consumer understanding of those technologies. She noted that survey results show that from 2014 to 2016 consumer understanding of lane-departure warning systems, collision avoidance–alert systems, blind spot monitoring–warning systems, and park assist–backup warning systems have all increased.

- Kolodge noted that blind spot warning and detection systems are one of the technologies most wanted by consumers and one of the most frequently used. Surveys indicate that 69% of consumers who have blind spot warning and detections systems use it every time they drive. She noted the importance of consumers having a positive initial experience for the ongoing use of technology.
- Kolodge discussed that the survey results show that even larger percentages of consumers reported wanting ADAS in their vehicles. Blind spot warning and detections systems were desired by 87% of the respondents, followed by park assist at 82%. Adaptive cruise control and low-speed collision avoidance were both desired by 71% of the respondents. She noted that these responses appear to be similar across age groups and vehicle preferences.
- Kolodge described the potential risks to consumer technology adoption due to confusion from industry variations. She noted that possible sources of confusion include the use of different marketing names, acronyms, icons, communication methods, and customization. There are also differences in default conditions, the ability to turn systems off, the last state remembrance, and operational characteristics.
- Kolodge used cruise control as an example of possible consumer confusion. She cited seven different marketing names in use, including adaptive cruise control (ACC), advanced smart cruise control (ASCC), automatic cruise control (also known as ACC), and active cruise control (also known as ACC). Other marketing names are cooperative adaptive cruise control (CACC), intelligent cruise control (ICC), and dynamic radar cruise control (DRCC). She further noted that differences in the operation of these systems may confuse consumers.
- Kolodge discussed market research on how consumers learn to operate different ADAS technologies. She noted that even with increasing market penetration, few consumers have prior experience with many ADAS technologies. Some consumers reported receiving instructions from the dealer when purchasing their vehicle. Reading the owner's manual and trial and error were other methods reported by consumers.
- Kolodge discussed the evolving consumer definition of quality. She noted that J.D. Powers has been studying quality for over 30 years. The initial customer focus on quality in the automotive industry was on the operation of the vehicle with no breakdowns. The second level addressed the ease or difficulty of use and understanding of new features. She noted the most recent level focuses on trust and relationship building with new technologies and systems. This level also addresses consumer comfort with the system accuracy, the match to expectations, and other related factors.
- Kolodge highlighted some of the responses to surveys J.D. Powers conducts of customers after 3 years of vehicle ownership. She noted that of consumers reporting problems with a blind spot monitoring system, 41% were false positives, 23% experienced false negatives, and 45% said the system did not work at all. She suggested that consumer trust will erode over time if these conditions continue.
- Kolodge noted that concerns surrounding increasing technology complexities, privacy, and the potential for systems to be hacked, hijacked, or crashed are prominent across all respondents. She reported that Generation Y and Z were nearly twice as likely as Generation X and five times as likely as Baby Boomers to fully trust automated self-driving technology. She noted that there is a relationship between consumer interest in full self-driving automation and their trust in the technology. She suggested that to achieve market penetration, full self-driving

automation should focus on the cohorts most accepting of it and allow the older generations more time and exposure to its benefits and market readiness.

- Kolodge discussed the importance of considering these factors in developing messaging for new automotive technology applications. She further noted that focusing on building and maintaining trust with products and services that are reliable and meet consumer expectations was important.

THE PUBLIC SAFETY CASE

Bryant Walker Smith, *University of South Carolina School of Law*

Bryant Walker Smith discussed potential liability concerns with automated vehicles. He described the public safety case as one approach for addressing liability. Smith covered the following topics in his presentation:

- Smith discussed the possible product liability for automation developers. He noted that if AVs are safer, the number of crashes may decline significantly. In this situation, he said that vehicle manufacturers may have a bigger slice of a smaller pie of total liability.
- Smith noted that possible liability issues were identified in a 1993 FHWA report: Advanced Vehicle Control Systems: Potential Tort Liability for Developers. He quoted the following text from the report. "The prospect of liability for catastrophic accidents resulting from a failure of [automated vehicle control systems] will likely deter entities from becoming involved with advanced vehicle control systems (AVCS) and impede its development unless the federal government adopts some or all the legislative [limits on liability]."
- Smith noted that two decades later this assertion is testable. Although the 1993 report's recommendations were not adopted, automakers have released many of the technologies identified and many companies are investing heavily in research and development. Further, he said that some companies have expressly accepted the current liability regime. For example, package shipping companies assume that some of their vehicles will be involved in crashes and incorporate the cost associated with those crashes into their fee structure.
- Smith noted that public expectations matter, as liability often depends on the perceived reasonableness of a company or its products. He suggested that if juries and judges think that a company has taken a reasonable approach they may be less likely to find that company liable.
- Smith described the public safety case to manage public expectations associated with AVs, as well as to obtain an approval or exemption. He noted that in this approach, a developer shares its safety philosophy with the public through data and analysis. The developer would explain how it defines safety, how it designs for safety, how it establishes safety, and how it monitors reasonable safety over the lifetime of a system. It is an opportunity for the developer to discuss the system's risk and opportunities, and to prepare the public for these systems.
- Smith noted that the public safety case can also be used to obtain an approval or exemption at the regulatory level. He described an example where a developer, as part of seeking a regulatory approval or exemption, makes a public argument for the safety of its system. The regulator, with input from the public, evaluates the reasonableness of that argument.

- Smith suggested that while regulators cannot have all the answers, they can get better at asking key questions. He further suggested that just as developers need space for technical innovation, regulators need space for regulatory innovation. He noted that the public is an essential partner in all these efforts and that public perception matters.
- Smith discussed if the public safety case could be viewed as an option or an obligation. He noted that many advanced vehicle technologies are consistent with state laws and that developers can choose to comply with or seek to change existing laws. He further noted that there are some regulatory gates and potential legal obstacles that may need to be addressed. Further, post-crash investigations will still occur and will add to the legal framework.
- Smith suggested that predicates for a public safety case included more expansive and explicit exemption authority at the federal and state levels, cultivation of technology-agnostic safety expertise at public agencies, and more robust public disclosure mechanisms. He predicted that when an automated driving (AD) developer shares its safety philosophy with the public through data and analysis, AD will be truly imminent.

HUMAN FACTORS RECOMMENDATIONS FOR HIGHLY AUTOMATED DRIVING IN THE EU PROJECT AdaptIVe

Marc Dziennus, *German Aerospace Center*

Marc Dziennus discussed the Automated Driving Applications and Technologies for Intelligent Vehicles (AdaptIVe) project and human-vehicle integration in automated vehicles. He covered the following topics in his presentation:

- The AdaptIVe project covers a 42-month time period from January 2014 to June 2017. The project budget is €25 million. There are 28 partners from the eight countries: France, Germany, Greece, Italy, the United Kingdom, Spain, Netherlands, and Sweden. Aria Etemad from the Volkswagen Group Research is the project coordinator.
- Elements included in AdaptIVe cover the SAEs levels of automation 1 through 4. Lane-keeping assistance and ACC represent Level 1. Parking assistance is a Level 2 partial automation feature. The traffic jam chauffeur is a Level 3 conditional automation feature. The parking garage pilot is a Level 4 high automation project.
- Dziennus noted that AdaptIVe includes the development of new functions on automobiles and trucks for AD, with research covering several scenarios. It also includes defining evaluation methodologies and addressing the legal framework.
- Dziennus described the human factors Sub Project 3 (SP3). Partners in SP3 include Volvo, Ford, the University of Leeds, Würzburg Institute for Traffic Sciences, and the German Aerospace Center. The main goals of SP3 are to support the partners with human factors knowledge and to homogenize development by providing human factor recommendations. The major work activities include creating use cases for the development and testing of automated functions, collecting existing human factors recommendations, and developing new research questions for human–vehicle integration. Other activities are conducting experiments based on the new research questions and creating new human factors recommendations. He described the general workflow that begins with the state-of-the art reviews and continues through developing research questions, conducting experiments, identifying human factors recommendations, and

developing demonstration vehicles. He noted that the focus is on human–machine interface (HMI) design for driver assistance in different automation levels, the transition between automated and manual driving, and system uncertainties and limits. The impact of system failure is also being examined.

- Dziennus reported that a total of 17 experiments have been conducted, which included surveys, simulator studies, and field studies. More than 300 drivers and 2,700 participants were involved in these activities. He summarized the results from the first round of experiments.
- The first experiment focused on the number of automation levels that should be displayed to the driver. The human factors recommendation was to use additional cues, such as sound and haptic cues, to indicate mode changes.
- The second experiment focused on the ability of drivers to resume control and respond in dangerous situations when they have been "out of the loop" in AD conditions. The human factors recommendation was to use visual feedback if available, such as a flashing steering wheel symbol, to indicate system uncertainty and to re-engage as the driver.
- The third experiment examined user preferences for a smartphone app for parking automation. The human factors recommendation was that short press and permanent interaction HMI modes are both possible.
- The fourth experiment examined the influence of a timely announcement of a traffic situation or driving behavior at take-over situations. The human factors recommendation was to provide timely information before approaching a situational change using visual and auditory feedback.
- The fifth experiment examined bringing a driver from Level 3 into Level 2 automation with peripheral cues and methods to help the driver anticipate automation behavior and failures. The experiment included installing a colored bar on the dashboard of the test vehicle. The human factors recommendation was to use a 360° ambient display for indicating detected vehicles to support drivers in anticipating automation maneuvers and failures.
- Dziennus noted that the final report for the project, which includes the final functional human factors recommendations, is under development and will be finalized by May 2017. He reported that the AdaptIVe final event will be held in Aachen, Germany, on June 28 and 29, 2017. More information on the project is available at www.AdaptIVe-ip.eu.

POLICY DEVELOPMENT AND AUTOMATED VEHICLES

Sarah Hunter, *X (formerly known as Google X)*

Sarah Hunter discussed possible policy implications associated with AVs. She suggested policy tools that would be of use in advancing the development and deployment of AVs. Hunter covered the following topics in her presentation:

- Hunter noted the diverse interest in self-driving cars, from the 70-year-old who needs to go to the doctor to the 7-year-old who wants to go to soccer practice. She discussed the policy implications associated with AVs. She noted that new regulations and laws are being considered in many states, as well as in countries worldwide addressing different aspects of driverless vehicles. These activities are occurring at a time when AVs are not being sold anywhere in the world. She commented that typically in transportation, innovation occurs first, followed by regulation.

- Hunter suggested that discussions between innovators and policy makers are crucial to the safe deployment of AVs. She noted that reducing crashes and improving safety was one of the key goals of AVs. She suggested that innovators and policy makers share this common goal.
- Hunter noted that just as innovators today are using different tools, policy makers also need different tools and new approaches, rather than using outdated tools. She described some of the elements needed in drafting new legislation, including industry standards and information about possible public risks. Since technology is not static and is evolving rapidly, she noted that policy considerations today may focus on technology that will no longer be in use when the laws and regulations are actually approved.
- Hunter suggested that laws and regulations should not hinder innovation. She further suggested that rather than writing new laws, developing new innovative policy-making tools would be more beneficial. She outlined three policy tools—government leadership, pilots, and public agency capacity building—that would be beneficial. She noted the government leadership in the U.K. as one example to consider in the United States. Leadership in the U.K. examined best practices and guidelines and brought together industry, innovators, and policy makers to reach consensus on an approach. She suggested the recent regulations in the United States on unmanned aerial vehicles (UAVs) provided another example of a good approach. She noted that these examples highlight government using its "soft power" of convening and facilitating.
- Conducting pilots was the second policy tool Hunter described. She noted the difference between testing and deploying self-driving vehicles. She suggested that using pilots to introduce small groups of consumers to AVs in selected areas was a good approach. She likened the process of AVs learning to navigate in an area to teaching a new driver. Both cases begin with simple situations and slowly progress to more complex environments. She noted that it is envisioned that self-driving vehicles (SDVs) will be introduced in areas with simple grid street systems and good weather. She noted that pilots provide great opportunities to gain real experience and better understand consumer behavior. She cited pilots in South Australia and Sweden, as well as the Smart Cities Challenge in the United States as good examples.
- The third tool Hunter suggested was capacity building at public agencies. She noted that regulatory and operating agencies would benefit from having staff with deeper expertise in technology. She commented that it is often difficult for public agencies to attract and retain technical staff, but that it would be of benefit to all groups to have additional technical expertise at public agencies.
- In closing, Hunter challenged symposium participants to consider 21st century policy tools for AVs. She noted that working together to identify creative and innovative policies will help advance the introduction of AVs that can save lives and change lives.

THE TRAFFIC JAM OF ROBOTS: IMPLICATIONS OF AUTONOMOUS VEHICLES ON TRIP MAKING

Joan Walker, *University of California, Berkeley*

Joan Walker discussed the potential impacts of AVs on travel, lifestyles, and trip making. She described research on these topics and areas for further investigation. Walker covered the following topics in her presentation:

- Walker noted there are numerous questions associated with how AVs might impact decisions on where to live, whether to buy a car or share one, trip making, and mode of travel. She commented that all of these decisions will in-turn influence traffic congestion levels and the location of congestion.
- Walker described one future vision with vehicles moving seamlessly on roadways and through intersections. This vision is in contrast to the heavily congested freeway corridors in most urban areas. She suggested that how this vision and current reality merge is a key question. In terms of the future reality, Walker suggested that automation will improve efficiency and safety, but not enough to relieve congestion due to powerful opposing trends. These trends include increasing population, increasing urbanization, and increasing vehicle miles traveled (VMT) per capita. She suggested that achieving the desired future requires discussion of automation and technology scenarios and actions.
- Walker discussed current travel behavior, noting that congestion exists because everyone wants to be at the same place at the same time and the network drive is constrained. She further noted that travel decisions are based on the need to get somewhere and tangible factors, such as travel time and cost, and intangible factors, such as convenience. Safety and contextual factors—social, emotional, and lifestyle—also influence travel behavior. She noted that a key question is how will automation influence these factors?
- Walker described simulation studies that have been conducted based on current and hypothesized travel behavior. Some studies suggest that AVs can drastically reduce the overall vehicle fleet by approximately 90%. This reduction could provide opportunities to repurpose parking structures and on-street parking. At the same time, these studies project increases in VMT.
- Walker discussed behavioral questions surrounding the introduction of AVs. These questions focus on the extent to which people will used AVs safely, share vehicles, share rides, and change their travel behavior. There are also questions if people will send their cars on errands and change their vehicle type. Increases in on-demand delivery needs may also occur. She suggested that research addressing these questions would be beneficial.
- Walker described some of the potential travel changes that may occur with the widespread use of AVs. One change is the ability of new demographics to travel by automobile, including younger and older people. Another change may be more travel because it is less onerous. Zero-occupancy trips are another possibility, and vehicle configurations may change.
- Walker noted that there is rich transport behavior literature that can be drawn on to help predict travel changes. Literature addresses pricing, multitasking, parking, sharing vehicles and rides, travel budgets, modal attitudes, habits, and social norms. She suggested that higher levels of automation will require new behavioral experiments, which are difficult as the technologies do not exist. Approaches to overcome this issue include simulation-based scenario

analysis, surveys with hypothetical scenarios, the use of virtual reality and gaming, and field experiments using analogous modes and prototypes.

- In planning for the future, Walker suggested not to underestimate emotional attachment people have to their vehicles. In moving toward shared vehicles, if the prices are affordable, signs lead to more vehicle miles. She further suggested the need to be proactive, not reactive. Once habits are formed, it is hard to change them. She noted that social norms are powerful and could be used to focus on desired behavior. She also noted that incentives may be needed. Other factors to consider are right sizing trucks for specific purposes, encouraging off-peak travel, discouraging zero-occupancy trips, and the evolving on-demand economy.

THE RIGHT ROLE FOR AUTONOMOUS VEHICLE TECHNOLOGY IN CITIES

Gabe Klein, *Fontinalis Partners and National Association of City Transportation Officials Strategic Advisory Board*

Seleta Reynolds, *Los Angeles Department of Transportation and President, National Association of City Transportation Officials*

Gabe Klein and Seleta Reynolds discussed issues associated with the current land use and development patterns and the potential for AVs to help address some of these concerns. They reviewed the National Association of City Transportation Officials (NACTO) policy statement on AVs. Klein and Reynolds covered the following topics in their presentation:

- Klein and Reynolds discussed issues associated with the automobile-centric society including traffic congestion, emissions and environmental concerns, crashes and fatalities, and urban sprawl. They described approaches to address these issues in cities, including redesigning streets around people not vehicles, complete streets, more extensive transit services, bike-sharing systems and bicycle facilities, and pedestrian facilities.
- Klein and Reynolds reviewed the establishment of the Interstate system and the suburban development it supported. They described recent trends in vehicle sharing, network services, and transit. They highlighted recent urban street, transit, and highway design guides that focus on people and that are changing historical practices. Examples of NACTO documents included the Urban Street Design Guide, the Urban Bikeway Design Guide, and the Bike Share Station Siting Guide.
- Klein and Reynolds discussed the complexity of urban areas and the benefits of the public and private sectors working together to address concerns. They noted that NACTO helps foster communication between the public and private sectors through conferences, workshops, training sessions, and other venues.
- Klein and Reynolds discussed the possible impacts of AVs on urban areas, including their potential to reshape cities. For example, the need for on-street parking and parking garages may be reduced. Autonomous electric transit vehicles of different sizes may be in operation, with Uber and Lyft as well as bike sharing providing first and last-mile connections. They noted that cities oriented around people, rather than vehicles, may be possible with an autonomous, shared, connected, and multimodal approaches. They discussed mobility-as-a-service, which reflects this approach and noted the importance of change management in moving to a new future.

- Klein and Reynolds reviewed the NACTO policy statement on AVs. They noted that NACTO supports a future transportation system that is sustainable, accessible, and affordable, supporting strong cities. New technologies, including fully automated vehicles, may support this vision, but may also have negative impacts. They noted that NACTO policies focus on promoting safety for all user groups, supporting the future vision of communities, rebalancing right-of-way use, supporting public transit, improving mobility for all demographic groups, and incentivizing shared, automated, and electric vehicles.

GENERAL SESSION

Keynote Session

MARK NORMAN
Director of Development and Strategic Initiatives,
Transportation Research Board, presider

TRANSPORTATION RESEARCH BOARD AND TRANSFORMATIONAL TECHNOLOGIES

Mark Norman, *Transportation Research Board*

Mark Norman discussed activities underway at TRB related to transformational transportation technologies. These technologies included CAVs, shared-use services, the next generation air traffic control system (NextGen), UAVs, smart cities, the Internet-of-things, big data, cybersecurity, alternative-fueled vehicles, and three-dimensional (3-D) printing. Norman covered the following in his presentation:

- Norman noted that *Transportation Research E-Circular 208: Transformational Technologies in Transportation: State of the Activities* was published in May 2016. It covers the research, conferences, outreach, and other efforts underway at TRB related to transformational technologies.
- Norman described the National Cooperative Highway Research Program (NCHRP) Project 20-102: Impacts of CVs and AVs on State and Local Transportation Agencies. The project was approved in December 2014, and $3.5 million has been authorized so far through mid-2016. Tasks under way include projects focusing on the implications of automation for motor vehicle codes, the impacts of CVs and AVs on regional transportation planning and modeling, dedicating lanes for use by CVs and AVs, and road markings for machine vision. Other tasks are developing evaluation guidance for AV pilots and demonstrations, examining the societal impacts of CV and AV systems, assessing impacts on transit operations, and CV and AV applications in truck freight operations.
- Norman highlighted the recently completed *NCHRP Legal Research Digest 69: A Look at the Legal Environment of Driverless Vehicles*. Topics addressed include civil liability for personal injury, criminal law and procedure, insurance, and privacy and security laws. He noted that TRB participated in a Congressional roundtable informing policy makers on research needs related to CAVs and transformational technologies.
- Norman noted that TRB is organizing the Transformational Technologies in Transportation Partners in Research Summit that will be held October 31–November 1, 2016, in Detroit, Michigan. The summit will bring together transportation policy and research leaders from public agencies, the private sector, and academia to explore partnering opportunities for identifying research needs and fostering research.
- Norman encouraged symposium participants to access the new TRB website www.TRB.org/main/TransTech.aspx for more information on transformational technologies.

The site includes TRB reports, academic reports, *Transportation Research Record* journal papers, TRB committee activities, research, conferences, and other information.

- Norman discussed Breakout Session 22: Can Our Research Processes Keep Up in an Age of Automated Vehicles and Other Transformational Technologies? Discussion topics for the session include possible options for more timely and strategic public-sector research, models from other sectors, and implementing new approaches.

WELCOME FROM THE TRANSPORTATION RESEARCH BOARD

Neil Pedersen, *Executive Director, Transportation Research Board*

Neil Pedersen provided a welcome from TRB. Pedersen noted that TRB was pleased to cosponsor the symposium with AUVSI. He also encouraged symposium attendees to participate in TRB activities, including committees and task forces, the annual meeting, and conferences and workshops. Pedersen introduced the keynote speaker, Mark Rosekind, Administrator, National Highway Traffic Safety Administration (NHSTA).

NATIONAL HIGHWAY TRAFFIC SAFETY ADMINISTRATION AND THE FUTURE OF AUTOMATED VEHICLES

Mark R. Rosekind, *National Highway Traffic Safety Administration*

Administrator Mark Rosekind discussed the work underway at NHTSA developing the Highly Automated Vehicle (HAV) Guidance. He described the questions the guidance is addressing and the general approach being taken. Rosekind covered the following topics in his presentation:

- Rosekind noted that it was an exciting time in transportation and an incredible period for the development of HAVs. He suggested that the future of HAVs was being created by symposium participants, as well as other individuals and groups throughout the world. As Secretary Foxx noted, Rosekind said that the U.S. DOT was forward leaning regarding AVs. He noted that reducing the 35,200 roadway fatalities that occurred in 2015 is a key reason for this forward-leaning focus. Further, he noted approximately 94% of crashes are due to human error. HAVs could potentially eliminate 19 of every 20 crashes. He noted that this life-saving promise of HAVs is at the center of the U.S. DOT's interest in HAVs. In addition, HAVs can enhance mobility for all user groups and provide other benefits.
- Rosekind discussed why the NHTSA was developing guidance at this time. He noted that NHTSA recently held two public meetings—one in Washington, D.C., and one in San Francisco, California—to help inform the development of the guidance. He said that the comments ranged from people thinking NHTSA was moving too quickly and guidelines would stifle development, to those thinking NHTSA was moving too slowly. Numerous comments also reflected the need for guidance and the timeliness of the current process. He suggested that industry needs some level of certainty as it makes major investments in HAVs. He further noted that various states are moving forward and are looking to NHTSA for guidance, and that various technologies are currently available in vehicles.

- Rosekind reviewed some of the questions that have been raised related to HAVs. A first question is "When will HAVs be safe enough?" He noted that this question is hard to define, much less answer. He suggested that technology should promote safety innovation. He noted that new technology should not introduce new safety risks, but that new tools are needed to save lives on the nation's roadways. He suggested that a thoughtful, but forceful approach is needed to bring life-saving technologies to the roadways. Rosekind identified related questions as "How safe is safe enough?" He suggested that HAVs should be much safer than current vehicles before widespread use will occur. He noted that new safety metrics will be needed for HAVs to address this question and to monitor HAV use. For example, a focus on crashes prevented and lives saved may be needed.
- A second question that many people ask is "What is even possible?" Rosekind noted that it is not the role of the federal government to select the technologies or to design the future. Rather it is government's role to establish a framework that will speed the development of technologies that have the potential to save lives.
- Rosekind noted that the traditional approach at the NHTSA is to set and enforce minimum standards. Recalls on automobiles not meeting standards reflect this traditional approach. He reported that OEMs have introduced new safety features, such as air bags, over the years. It is only after these features have been in place for a number of years that the NHTSA develops guidance. He noted that this approach has yielded important safety benefits, but it also has limitations. One limitation in the length of time it takes to move guidance through the process, which may take 6 to 10 years. He suggested that this traditional approach may not meet the needs of the rapid technology advancements experienced today. He stressed that strong safety guidance will continue to be a key part of NHTSA's mission.
- Rosekind noted that the approach being taken with the HAV guidance is intended to be nimble and flexible to allow for keeping pace with technological change. It will also outline more-specific guidelines on the introduction of new technology. He noted the guidance is an important first step to lead to a more harmonized approach, but that it is only a first step. It will help provide more certainty to manufacturers and developers and will provide clear expectations on safety.
- In closing, Rosekind commented that vehicle safety features have saved an estimated 613,501 lives from 1960 to 2012. He noted that HAVs hold the promise to save even more lives.

International Experience

DAVID AGNEW
Director of Advanced Engineering, Hyundai Mobis North America and AUVSI Board of Directors, presider

CONNECTED AND AUTONOMOUS VEHICLES IN THE UNITED KINGDOM

Iain Forbes, *Department for Transportation, United Kingdom*

Iain Forbes described AV projects and research underway in the U.K. He discussed some of the lessons learned from these activities. Forbes covered the following topics in his presentation:

- Forbes noted that the U.K. faces the same transport issues and challenges as the United States and countries throughout the world and sees the same benefits from AVs. He focused on some of the key lessons learned in developing and deploying advanced transport technologies in the U.K., including working together, solving the problems confronting you first, keeping learning, and deriving as many benefits along the way as you can.
- Forbes discussed approaches used to foster working together first. He noted that over the next few years approximately $.5 billion will be spend in Britain for research and development related to CAVs. A year ago he highlighted projected underway including UK AUTODRIVE in Milton Keys and Coventry, VENTURER in Bristol, and Gateway in London. He noted that many other projects have been initiated in the past year. Numerous public and private groups are participating in these projects.
- Forbes described the O2 Gateway Project in London, which is an autonomous electric shuttle. Groups involved in the project include software companies, manufacturing businesses, research companies, the Greenwich Council, the Royal College of Art, and insurance companies. He noted that this broad range of partners, who have interest in different aspects of the project, are working together to ensure its success. He also highlighted the groups involved in other projects and stressed the importance of collaboration.
- Forbes discussed the second lesson learned of solving the problem in front of you first. He noted that some problems can be solved through regulations and rules, some problems do not need regulations and rules, and some problems can be created through regulations and rules. He suggested that the approach in the U.K. is to focus on the first, ignore the second, and stay away from the third.
- Forbes noted that the U.K.'s Code of Practice for Testing Autonomous Cars was announced at the 2015 AVS. He suggested that this document make the U.K. one of the most open environments for developing and testing AVs. The code of practice was developed with industry input. It did not involve changing the regulatory framework. He noted a second example of a recently published review of regulations for AV technologies. He noted that in the U.K., the driver, not the vehicle is insured. He suggested that this approach would make it difficult to deal with a crash involving a driverless AV. As a result, the insurance framework in the U.K. is being updated to address AVs and other related activities. Changes are also being made to traffic

guidelines to maximize benefits from automated and autonomous technologies. These are examples of problems that need to be solved now.

- The third lesson Forbes described was to keep on learning. He noted that a key question is "What will the impact of autonomous vehicles be on the future transport system in the U.K.?" This is an important question as decisions on future transport infrastructure are being made now. Forbes presented two videos of using microsimulation to model the shorter headways of AVs and bunching at intersections.
- Forbes also noted the importance of understanding public perception and behavior, which will be key to market acceptance of AVs. He described a 3-year program conducted by the University College of London assessing public perception.
- The fourth lesson described by Forbes was to derive benefits from projects along the way. He suggested that numerous benefits can be realized along the pathway to AVs. He highlighted one example of a company building on their computer game expertise to model communities.

EUROPEAN ACTIVITIES ON CONNECTED AND AUTOMATED DRIVING: THE PRESENT AND BEYOND—AdaptIVe and AUTONET2030 USE CASES

Angelos Amditis, *Institute of Communication and Computer Systems*

Angelos Amditis described examples of automated and connected driving activities underway in Europe. He highlighted the development and status of the AdaptIVe and AUTONET2030 projects, and discussed the future holistic view of automation. Amditis covered the following topics in his presentation:

- Amditis noted that intelligent transportation systems (ITS) has been a major focus in Europe to improve the movement of people and goods for the past 20 years. Recent activities address control strategies, vehicle-to-everything (V2X) connectivity, human factors, and automated functions.
- AdaptIVe is an EU-funded project covering 42 months from January 2014 to June 2017. The project includes eight countries and 28 public and private-sector partners. France, Germany, Greece, Italy, Spain, Sweden, Netherlands, and U.K. are the participating countries. The Volkswagen Group is the project coordinator.
- Amditis reviewed the high-level project objectives, which include supporting drivers in demanding or repetitive tasks to make travel more comfortable. Other objectives are to dynamically adapt vehicles to the level of automation according to the current situation, to develop vehicles that react more effectively to external threats, and to design vehicles that are resilient to different types of system and human failure.
- AdaptIVe includes highway scenarios, urban scenarios, and close-distance scenarios. These scenarios may use V2X communication, collaborate control, and redundancy to minimize risk. There are also three horizontal subprojects focusing on legal issues, human factors, and evaluation.
- The close-distance scenarios include application-oriented subprojects addressing parking. Amditis noted that one effort is using groundtruth detailed maps of parking spaces, along with radar-based grid maps, data clustering, vehicle classification, and cognitive mapping.

Examples of highway scenarios include cooperative merging on highways using V2V communications and redundancy to safety stop level 4 vehicles if needed.

- Amditis reviewed some of the legal issues being examined including liability and road traffic and regulatory laws. Data security and data privacy are also being examined. Further, automotive standards are being reviewed. He noted there is a need to develop a new code of practice, as the Code of Practice for ADAS covers the needs for safety validation only partially for AD functions. The capability of different sensors under a variety of conditions has also been tested.
- Amditis noted that the human factors activities are examining the transitions of control between automation and driver. He noted that as long as there are not fully autonomous systems, systems will always have to interact with humans at different times and to different degrees. The goal is safe and efficient transitions. He noted that 17 simulator studies involving more than 300 participants and drivers and one survey of 2,700 individuals have been conducted. A public catalog with human factors recommendations for AVs should be available in June 2017.
- Amditis described AUTONET2030, which focuses on enabling the convergence of pure sensor-based automation with cooperative V2X communications and decentralized maneuvering control algorithms. It includes a consortium of nine partners.
- Amditis discussed the three main AutoNet2030 research threads. The first research thread focuses on decentralized cooperative maneuvering control algorithms to enhance automated maneuvers using mutual information sharing. It includes automotive requirements for cooperative controls and decentralized decision-making algorithms for lane changing–merging and intersection management. The second research thread addresses specification and standardization of V2X communication protocols for AD to achieve a fast and reliable exchange of maneuvering data. The third research thread examines onboard architecture for integrated sensing and HMI-based adviser maneuvering to deploy a maneuvering system for automated (manually driven) vehicles.
- Amditis discussed the holistic view on automated transportation systems. He noted that most efforts in automated transport focus on the vehicle side and that no matter how intelligent or automated a vehicle is, it might still cause other problems, such as congestion and incidents. He suggested that automation is needed in other elements of the transport system, including infrastructure, traffic management centers (TMCs), and vulnerable road users (VRU).
- Amditis noted that the transportation infrastructure will play an especially important role to support the transition period which includes mixed traffic with legacy vehicles and different types of equipped vehicles. He described the digitalization of the road infrastructure, noting that highly accurate digital maps, dynamic information from AV sensors and infrastructure sensors, and advanced communication and positioning technologies will all be important. He also noted that physical infrastructure adaptations and upgrades will also be needed.
- In concluding, Amditis reviewed some of the key findings from the various projects. He noted that parallel development of technology, legal, and operational aspects is required and that different levels of automation can be applied in different application areas. He commented that vehicle–human interaction will always be necessary even in the higher levels of automation. He noted the need for a code of practice and standardization/certification for automated functions and the need for extensive standardization efforts to address connectivity issues. He suggested that strategies will be affected by automation and that mixed traffic with automated and

nonautomated vehicles will create unexpected situations. The infrastructure will play a key role during this period. He further suggested that a holistic traffic and transport approach with collaboration among all groups is needed to accomplish a new mobility paradigm for people and freight.

CONNECTED AND AUTOMATED DRIVING IN THE NETHERLANDS: CHALLENGE, EXPERIENCE, AND THE DECLARATION OF AMSTERDAM

Tom Alkim, *Ministry of Infrastructure and the Environment, Netherlands*

Tom Alkim discussed CAV activities in the Netherlands, including the Declaration of Amsterdam and the EU Truck Platooning Challenge. He presented a video to highlight some of these activities. Alkim covered the following topics in his presentation:

- Alkim noted that the Netherlands had the EU Presidency in 2016, which was used to highlight smart mobility. The Declaration of Amsterdam provides a strategic approach to the Cooperative Intelligent Transport System (C-ITS) Platform and AD. AD was showcased by providing Ministers with experiences to AVs and through the EU Truck Platooning Challenge.
- Alkim described the EU Truck Platooning Challenge. The challenge was organized to demonstrate that platooning is technically feasible, but that there are policy differences among countries that must be addressed before widespread use of truck platooning will be possible. Six European truck manufacturers participated in the challenge. Platoons of trucks traveled from five countries to the Netherlands as part of the challenge. More information on the challenge is available at www.eutruckplatooning.com.
- Alkim noted that a summary of the Truck Platooning Challenge was developed. The report summarizes the exemptions needed in the different countries, interviews conducted with truck drivers, and analyses of aerial films of interaction with other traffic. The report is available at www.tentdays.eu/2016.
- Alkim described the experience of the Ministers driving partly AVs in Amsterdam. A total of 14 vehicles from different automotive companies were used. He noted that the event generated a lot of media coverage, as well as positive feedback from the Ministers who operated the vehicles.
- The Ministers of Transportation from the EU member states also met in Amsterdam as part of the AV experience. The Ministers signed the Declaration of Amsterdam, which addresses cooperation in connected and AD. It is available at https://english.eu2016.nl/documents/publications/2016/04/14/declaration-of-amsterdam. He noted that the Declaration addresses the technical and policy challenges of moving forward with AD in the EU. It was drafted with the European Commission (EC), the member states, and industry. It includes joint actions and activities individual states can undertake. He noted that connected, cooperative, and AD developments should come together to bring societal benefits.
- Alkim highlighted Welcome to Knowledgeagenda Automated Driving, which are online overviews of available and required knowledge in the field of AD and self-driving vehicles. The website includes a library with reports, papers, and presentations, as well as information on events and tests. It also includes videos and frequently asked questions. The website is http://knowledgeagenda.connekt.nl/engels/.

- Alkim described the Internet Consultation on Experimenting Law. He noted that the Dutch government wants to facilitate experiments without a driver in a vehicle. Creating a legal framework, based upon exemptions, to allow for testing is envisioned to be the first step. Permanent rulemaking would follow based on the results of these tests and other research. The Internet consultation on the proposed draft for such experiments will be open until September 1, 2016. Input and comments from road operators, road users, parties conducting experiments, and other stakeholders is being solicited.
- In closing, Alkim reported that earlier in the week a Level 4 self-driving bus was piloted using a bus-only lane in the Netherlands. More information on the Mercedes-Benz Future Bus with CityPilot is available at www.daimler.com/innovation/autonomous-driving/future-bus.html.

i-GAME: FROM PLATOONING TO COOPERATIVE AUTOMATED MANEUVERING

Jeroen Ploeg, *TNO Automotive, Netherlands*

Jeroen Ploeg discussed the i-GAME project including the EU-funded Grand Cooperative Driving Challenge (GCDC) AutoMation Experience. He covered the following topics in his presentation:

- The i-GAME objective is to speed-up real-life implementation and interoperability of cooperative AD driving through both development and demonstration. The development of interaction protocols is a key element, as is creating and growing an expert community. He cited an example using real-life vehicle merging to define a merging scenario, which is used to develop the interaction protocol.
- Ploeg noted that in 2011, the GCDC focused on vehicle platooning. The GCDC is going beyond platooning in 2016 to address complex traffic scenarios, including merging, intersections, and lane reductions.
- Ploeg described two i-GAME simulated automated maneuvering scenarios: merging on highways and cooperative intersections. The challenge with merging on a highway is to negotiate and perform the merge before the lane gap is closed. The challenge of the cooperative intersection is to optimize the passage of an oncoming vehicle at a T-intersection.
- Ploeg noted that to obtain a generic solution, the scenarios are decomposed into maneuvers. The maneuvers are executed by a set of high-level controllers, also known as agents. The scenario execution is managed by the interaction protocol, which triggers certain maneuvers. The merging scenario control includes basically two longitudinal controllers: vehicle following using CACC and obstacle-avoidance control. It also includes a lateral controller to perform the lane change.
- In describing the intersection scenario control, Ploeg noted that the basic approach transforms the intersection problem into a virtual platooning problem, avoiding collisions by assigning the correct vehicle order in the virtual platoon. Ploeg described the wireless communications, which is based on the ITS-G5 protocol stack.
- Ploeg noted that 10 student teams from six European countries participated in the 2016 GCDC, which was held during the Dutch Technology Week in May. The teams were judged on a number of criteria, with the Swedish team from Halmstad University winning first prize.

CityMobil2: FOUR YEARS OF DEMONSTRATING AUTOMATED ROAD TRANSPORT SYSTEMS IN EUROPEAN CITIES

Adriano Alessandrini, *Università degli Studi di Firenze*

Adriano Alessandrini discussed CityMobil2, which involves demonstrating automated road passenger transport systems in European cities. He described the pilots sponsored by the EU in seven cities, the results to date, and future activities. Alessandrini covered the following topics in his presentation:

- Alessandrini described the elements of CityMobil2. In addition to demonstrating Automated Road Transport Systems (ARTS), CityMobil2 proposed a safety assessment procedure that can be applied in national legal frameworks for certification. It also is examining the long-term socioeconomic effects of different automation scenarios.
- CityMobil2 included large-scale demonstrations, small-scale demonstrations, and showcases. Large-scale demonstrations of driverless buses were conducted in La Rochelle, France; Lausanne, Switzerland; and Trikala, Greece. Small-scale demonstration sites included Oristano, Italy; Vantaa, Finland; San Sebastian, Spain; and Antibes, France. The demonstrations included different operating characteristics, including removing on-street parking, shared right-of-way, and protected right-of-way. The demonstrations served more than 60,000 passengers.
- Alessandrini described some of the results from the demonstrations. He noted that users are enthusiastic about the systems. Further, additional cities are ready to commit to demonstrating and deploying self-driving shuttles. Four EU member states now have regulations allowing trial deployment. He noted that cities need to be willing to invest political capital for successful projects and be ready to provide adequate space for safe operation at higher speeds.
- Alessandrini highlighted the results from user surveys conducted as part of the demonstrations. Over 80% of the respondents favored permanent implementation of the service on the existing route or on routes serving other parts of the city. However, few riders were willing to pay more for the service. Many users responded that the fares should be the same as fares for existing transit services.
- Alessandrini discussed the perceptions of pedestrians and bicyclists to the safety of the systems, as measured by responses to photographs with and without road markings. He noted that respondents reported feeling safer with the road markings. They also reported that the lane markings provide an indication of priority for the ARTS vehicles.
- Alessandrini summarized future activities. He noted that at least four cities are moving forward with permanent implementations. New proposals to the EC are also being prepared to demonstrate how automation can solve more than just the first- and last-mile link. He noted that the autoKAB company was created to retrofit existing busses into self-driving buses. He noted that existing transit operators may need to change their business models to accommodate self-driving vehicles. He reported that a new ad hoc integrated service provider company, called MEDIUM, has been created to help facilitate this change.

DRIVE SWEDEN: A NATIONAL EFFORT FOR AN AUTOMATED TRANSPORTATION SYSTEM

Jan Hellåker, *Lindholmen Science Park AB*

Jan Hellåker discussed the Drive Sweden Strategic Information Program. He described key elements of the program, public and private sector participants, and future activities. Hellåker covered the following topics in his presentation:

- Hellåker noted that Drive Sweden is a long-term program focused on developing and piloting the next-generation mobility system for people and goods. Launched in 2015, it currently includes four program phases through 2027. The focus is on mobility and service through shared CAVs. He noted that the service must be inclusive of all segments of the population and all transportation modes.
- Hellåker highlighted the Drive Sweden partners, which include government, academia, and industry. Government agencies at the national and local levels, universities, automotive and electronic companies, and other businesses are involved. He noted that collaboration is a key element of the partnerships.
- Hellåker described some of the current projects and activities that Drive Sweden is building upon. One example was the Drive Me SDVs for sustainable mobility project in Göteborg was initiated in 2013. It is expanding to 100 families using SAE Level 4 vehicles. A second example was Volvo and Scania participants in the EU truck platooning demonstration. He noted that off-road automation was an important area, with applications in the mining and forestry industries, as well as refuse collection.
- Hellåker noted that the electric vehicle maker NEVS was moving forward with a focus on shared mobility. NEVS' vision is to shape mobility on product, service, and system levels. He also described the Wallenberg Autonomous Systems Program (WASP), which is a $200 million, 10-year research and development effort. It includes both strategic and thematic areas, research laboratories, and demonstrations.
- Hellåker described ASTAZERO, a new full-scale test facility for tomorrow's active safety and automation. The facility covers 2,000 acres and includes a rural road, a multilane road, a high-speed area, a four-block city area, and a proving ground. The facility also has its own, dedicated 4G/LTE cellular network.
- Hellåker described the Kista Mobility Week, which included the use of two CityMobil2 self-driving shuttles operating between a subway station and employment locations. The demonstration was well received, with 3,300 people riding the self-driving shuttles over a 5-day period.
- Hellåker discussed the development of policy addressing self-driving vehicle trials. To obtain a trial permit, the test organization has to identify how road safety will be ensured when the trials are occurring. The Swedish Transport Agency is responsible for authorizing permits to conduct trials.
- Hellåker summarized the 2016 priorities focusing on mobility services for people and goods. He noted that developing a common platform for cloud services for use with traffic management of CVs was a major priority. Examining methods and tools for conducting impact assessments is another priority.

- In concluding, Hellåker noted that Drive Sweden is more than just driverless vehicles. It is a completely new approach to mobility. He suggested that a shift is occurring and that new mobility business models will emerge enabling sustainable cities.

GENERAL SESSION

Key Enablers

Standards, Safety, and Cybersecurity

RICHARD BISHOP
Bishop Consulting, presider

AUTOMATED AND CONNECTED TRANSPORTATION STANDARDS: KEY TO NEW VEHICLE TECHNOLOGIES

Jack Pokrzywa, *SAE Global Ground Vehicle Standards and US TAG ISO TC22*

Jack Pokrzywa discussed the SAE automated and connected transportation standards. He described the background and development of the standards and highlighted examples of recent standards related to vehicle technologies. Pokrzywa covered the following topics in his presentation:

- Pokrzywa noted that in 2009 SAE was given the responsibility by the U.S. Department of Energy (DOE) for developing standards for hybrid plug-in electric vehicles (PHEVs). SAE worked with stakeholders to create interoperable standards between PHEVs and the smart grid and develop a roadmap for PHEVs.
- Pokrzywa reported that 55 SAE committees are working across the passenger and commercial vehicle sectors on different automated and connected transportation standards. The 46 SAE standards address terms and conditions, security, interoperability, vehicle and system requirements, privacy, and safety.
- Part 1 addresses terms and definitions, interoperability, and vehicle and system performance requirements. Under the terms and definitions, he noted that the SAEJ3016-Taxonomy and Definitions for Terms Related to On-Road Motor Vehicle Automated Driving Systems was adopted in 2014. He noted that Version 2 was just adopted, which clarifies the taxonomy for lower automation levels, modifies existing terms and definitions, adds new terms and definitions, and adds examples and explanatory text. He noted that one of the key standards in Part 1: Interoperability is SAE J2725, Dedicated Short Range Communication (DSRC) Message Set Dictionary.
- Part 1 Vehicle and System Performance Requirements includes DSRC common performance requirements (standard in progress, SAE J2945), on-board system requirements for V2V safety communications (SAE J2945/1), performance requirements for cooperative adaptive cruise control and platooning (SAE J2945/6), and performance requirements for safety communications to VRUs (standard in progress, SAE J2945/9). SAEJ3131, AD reference architecture, is a standard under development.
- Pokrzywa noted that there are a number of standards in Part 2: Safety that focus on the areas of functional safety, safety and reliability, active safety, safety and human factors, and other safety. He said that SAE J3018: Guidelines for Safe On-Road Testing of SAE Levels 3, 4, and 5 Prototype AD Systems was a much requested guide. Uniform Pavement Markings for Machine Vision Systems is a standard in progress as part of a coordinated effort between SAE,

the American Association of State Highway and Transportation Officials (AASHTO), and TRB. He also noted that SAE J3088: Advanced Driver Assistance Systems and Levels of Automation is a standard in progress.

- Part 3 includes privacy and security. Pokrzywa noted that SAE J3061: Cybersecurity Recommended Practice for Cyber-Physical Vehicle Systems is an industry first. Cybersecurity cuts across all the automated and connected areas. He reported that the potential of making J3061 a joint standard with ISO is being discussed. SAE J3101: Requirements for Hardware Protected Security for Ground Vehicle Applications is a standard in progress. SAE J3005: Permanent or Semi-Permanent Installed Diagnostic Communication Devices provides guidance to vendors who are plugging into on-board diagnostic (OBD) ports.
- In closing, Pokrzywa invited others interested in standards to work together to develop a roadmap to guide further developments.

CYBERSECURITY CHALLENGES FOR AUTOMATED VEHICLES

Jonathan Petit, *Security Innovation, Inc.*

Jonathan Petit discussed technical and governance challenges associated with cybersecurity for AVs. He also suggested possible activities to begin addressing these challenges. Petit covered the following topics in his presentation.

- The first technical challenge discussed by Petit was associated with sensing technologies, which produce raw data. The onboard unit processes and analyses the data. Cloud services are used for processing and storage. He noted that data in the infrastructure is also a key element. Possible constraints identified by Petit were cost, computation overhead, space, energy, and communication overhead. Possible challenges he identified were hardware security, software security, operating system security, network security, and privacy.
- Examples of technical challenges associated with sensors cited by Petit included pen testing or assessing the security level and protection needed. He suggested that learning from military applications was a good approach. Petit noted that updating the software and firmware over-the-air is critical. He noted that it might be possible to make sensors smarter so that they can play an active role in the defense system.
- Petit described technical challenges associated with processing, filtering, and normalization of the data obtained by the sensors. A typical challenge focuses on hardware host-based security with minimal overhead. He noted that side-channel protection was possible. He also noted that middleware security, which may be used to process that data is an issue. Other challenges were sensor authentication and in-vehicle network security.
- Sensor function technical challenges described by Petit included hardware security with the central processing units (CPUs). Methods to harden adversarial input, and differentiating attacks from faulty sensors. He also noted that sensor fusion may use external V2X, map, and cloud service data. Challenges with external data included the extent of trust in these sources, consensus in V2X network, the integrity and freshness of maps, and the security of cloud services.
- The final technical challenges described by Petit focused on the action a vehicle takes based on the data. Deciding what is best for the vehicle and the community is a challenge. Petit

suggested that building a context-aware misbehavior engine that includes prediction, detection, and reaction as we approach to addressing this challenge.

- Petit discussed possible governance challenges. A first governance challenge he identified was encouraging security by design and security testing. Examples of three approaches include focusing on upgradability, crypto agility, and security ratings. He noted that another governance challenge is that data ownership will affect privacy and security. Another governance challenge he noted related to having a forensics ability system in place to identify misbehaving vehicles and other issues. He also suggested that the role of road operators may change and that new policies may be needed to address these changes.
- Petit reviewed the suggestions for addressing some of the technical challenges including pen-testing sensors and hardening them, sing hardware/host-based security, and using a security level as weight. Other approaches included securing external (contextual) data and developing misbehavior detection capabilities.
- Petit suggested that organizing and conducting workshops with subject matter experts to establish cybersecurity guidelines for AV would be an appropriate next step, including hardware, software, network, and privacy.

SAFETY ASSURANCE FOR HIGHLY AUTOMATED DRIVING: THE PEGASUS APPROACH

Hermann Winner, *Technische Universität Darmstadt*

Herman Winner discussed the safety assurance of highly automated driving (HAD). He presented a possible approach for considering the safety impacts of AD. Winner covered the following topics in his presentation:

- Winner noted that HAD fits with the SAE Level 3 with no responsibility of human drivers (operators) during operation of automation, but the automation may shift the driving task back toward humans in a reasonable transition time. He commented that HAD is expected as an introductory path to fully automated or driverless driving. He cited a typical use case as the Autobahn Chauffeur, which has a maximum speed of 130 km/h. He noted that function availability depends on preconditions being met. He noted that HAD has an advantage over Level 4 systems in that it can rely on the capability of humans for the handling of unknown or complex situations. A possible issue is that the transition might lead to new risks.
- Winner suggested that progress in safety by automation has to be measured in comparison with today's risk, with at least two relevant categories addressed. These two categories are crashes with damage to persons and crashes with fatalities. He described some of the differences between conventional and AVs, which could influence the safety of HAD.
- Winner discussed available information on driving safety performance. He noted that data on crashes and their causes is available from different sources. Information on driving safety—such as exceeding the speed limit—may be available from these sources. He also noted that there are qualitative models for driver information processing and driving tasks, as well as for simple scenarios. There are also human-reliability models that can be used to interpret observed accident frequency. He described the Swiss Cheese Model to develop a simple probabilistic accident model. He noted that an accident usually involves a number of factors. He

noted that there is a lack of understanding and information on the frequency and type of non-standard situations, either self-caused or innocently exposed, and the performance of human drivers in nonstandard situations. There is also not a good understanding of the accident avoidance capabilities of human drivers. He suggested that a dark matter problem exists, as information is only available on standard scenarios and the reported accidents. The probability for transition from accident-free driving to real accident occurrence is not known. He suggested that avoiding the known human accident causes are not sufficient for considering HAD. He noted that the accidents avoidance capability of humans is not recorded. There are also no quantitative figures about types of critical scenarios and their frequency where humans avoid accidents. One approach he suggested was to begin by considering the uncritical scenarios with very low potential for accidents, the critical scenarios, with potential for accidents, and the true accident scenarios.

- Winner presented an accident model for AVs, which adapts the Swiss Cheese Model for AD. In addition to the three previous uncritical, critical, and true accident scenarios, it includes new critical scenarios based on automated risk exposure and a new type of automated accidents.
- Winner noted that the functional design of AD promises higher safety by the reduction of frequency of known critical situations. He suggested that unknowns include the capability of AD to avoid accidents in the remaining critical situations, and the frequency of new critical situations generated by AD and the capability to control them safely. He commented that the validation of AD has to cover both and has to gain all necessary knowledge prerequisites.
- Winner described PEGASUS, which is the Project for Establishing Generally Accepted quality criteria, tools, and methods, as well as Scenarios and (in German: Und) Situations for the release of highly automated functions. It was founded by the German Federal Ministry for Economic Affairs and Energy (BMWi). He noted that PEGASUS is intended to close gaps in the area of testing and approving AVs with the aim of transferring existing highly automated vehicle prototypes into products. PEGASUS includes the five German OEMs, as well as Tier 1 suppliers, test laboratories, scientific institutes, subject matter experts, and subcontractors. He described examples of prototypes, test laboratories, and products. He identified the following questions being examined. What performance and safety criteria do systems for HAD need to fulfill? How do we validate their performance?
- In conclusion, Winner noted that PEGASUS is a national project implementation for fast progress in AD. He said that embedding knowledge into industry, as well as dissemination of knowledge and experience across the appropriate committees for standardization is an important focus. He also noted that there is open access to all essential project results, and that collaboration with other consortia and safety experts is highly appreciated.

PANEL: STARTING UP A TRANSPORTATION REVOLUTION

Bob Denaro, *TRB Joint Subcommittee on Challenges and Opportunities for Road Vehicle Automation and ITS Consulting; moderator*

Panelists: Louay Eldada, Quanergy Systems, Inc.; Nalin Gupta, Auro Robotics; Sravan Puttagunta, Civil Maps

This panel featured discussions with the CEOs of three technology startup companies: Louay Eldada, Nalin Gupta, and Sravan Puttagunta. The panelists responded to questions from the moderator and the audience. The three panelists covered the following topics in their remarks:

- The products made by the three companies included LiDAR and related software, software SDA that enables AVs to drive anywhere, and autonomous shuttles for campuses. The panelists noted that their business fill specific needs within the AV and technology space. They also noted that start-up companies can afford to take risks and that the culture and creativity of start-ups is attractive to investors and to the OEMs who may use their products and services.
- The panelists discussed regulatory and policy concerns related to AVs and the possible impacts on start-up companies. Providing information to decision makers on the benefits of different technologies and products, as well as documenting the safe operations, was suggested as important. The need for regulations or certification for ensuring data quality was also suggested as important by some participants.
- Questions associated with data ownership and data privacy were discussed by the panelists.
- Panelists discussed the need for redundancy in technologies and the advantages and limitations of different technologies for various applications.
- The global market for AVs and related systems was explored by the panelists. China was noted as a growing market.
- Participants highlighted the intellectual and creative energy of personnel at start-ups. They noted the current talent gold rush and the demand for software developers and for individuals with nonautomotive skill sets. At the same time, participants suggested the important benefits of mentoring from experienced leaders at established businesses. It was suggested that start-ups are pushing OEMs out of their comfort zone.
- Participants described some of the factors that might slow down the introduction of AVs. Examples of possible limitations included vehicle functionality to navigate and make decisions based on location and position and the state-of-the-infrastructure, translating the human context into vehicle capabilities, and addressing the "corner" cases or the 1% of the cases that will take the most time.

Update from Europe, Japan, and U.S. Federal Agencies

DAVID AGNEW
Hyundai Mobis North America and
Member, AUVSI Board of Directors, presider

EUROPEAN COLLABORATION ON ROAD AUTOMATION

Liam Breslin, *Directorate–General Research and Innovation, European Commission*

Liam Breslin described the organization of CAV policy development, research, testing, and deployment in Europe, including activities supported by the EC. He noted the widespread interest in road automation and the numerous testing activities underway. Breslin covered the following topics in his presentation:

- A number of tests have been conducted in Europe focusing on different aspects of AD and more are being planned. These tests involve passenger vehicles, automated urban mobility systems, and cooperative driving.
- The 2016 European Truck Platooning Challenge was the first successful experiment with cross-border platooning. It included the testing of semi-automated trucks in platoons on public roads from five European cities to the Port of Rotterdam in the Netherlands. The six major truck manufacturers in Europe participated, along with the five member states. A key element of the challenge success was the teamwork between the public and private sectors, including road and vehicle authorities, logistics services companies, universities, and stakeholder bodies.
- Breslin noted that connected and automated road transport is expected to shape the future of road transport and mobility in Europe. The rapid development of connected and AD technologies is expected to continue, but there are still concerns that need to be addressed.
- There are a number of EC initiatives on connected and AD. The Strategic Transport Research and Innovation Agenda (STRIA) initiative will support the development and deployment of transport low-carbon technology solutions for all modes. STRIA develops roadmaps to define research and innovation options for seven technical areas of electromobility, alternative fuels, vehicle design and manufacturing, connected and automated transport, transport infrastructure, network and traffic management systems, and smart transport and mobility services.
- The objectives for the roadmap on connected and automated transport include identifying potential contributions to achieving the EU climate and energy competiveness goals, developing an integrated long-term transport research and implementation (R&I) strategy on connected and automated transport, and defining policy options supporting research, innovation, and wide market uptake for each thematic area. The roadmap addresses all transport modes, including air, rail, road, and water.
- A second EC initiative is the high-level group of the competiveness and sustainable growth of the automotive industry in the EU (GEAR 2030). It focuses on supporting the EU automotive industry to strengthen its competiveness and to address the new challenges it is

facing in a coordinated manner. There are working groups examining the competitiveness of the European automotive industry, the automotive value chain, and AVCS. The working group on highly AVCS will develop a roadmap on connected and AD, propose actions to achieve a fast roll out of connected and AD technologies, and examine financing tools and legal and policy issues.

- Another EC initiative is the C-ITS platform. This initiative, which has been underway for a few years, will provide a platform for the development of cooperative and intelligent transport systems. It will provide conclusions and recommendations that are relevant for road automation in the area of data protection, cybersecurity, and access to in-vehicle data, digital infrastructure, traffic management, and road safety.
- The Oettinger Roundtable on connected and AD is focusing on the digital aspects of connected and AD, including network coverage, standards, interoperability, and cybersecurity. It is bringing together the automotive and the telecommunication industries to develop a proposal for a pan-European, large-scale project that will provide enabling connectivity.
- The EC has sponsored research on automated road transport for a number of years, with a project on driver-assistance systems initiated in 2005. Research focusing on automated applications is more recent, however. Research involves partnerships among multiple private- and public-sector groups, including automotive companies, equipment suppliers, member states, and universities. The intent is not to duplicate efforts of the automotive industry or individual countries. Rather, the intent is to focus on topics that are of mutual interest to member states. Examples of current EC R&I priorities include user and social acceptance, road infrastructure needs, and the safety of advanced driver systems in complex traffic situations.
- The automated road transport topics in 2016–2017 focus on supporting the short-term introduction of AD systems for passenger automobiles, trucks, urban transport, and large-scale field operational tests of technologies in complex traffic and driving conditions. A total budget of €114 million was allocated for projects in these areas. Large automation pilots for passenger vehicles involving multiple automotive manufacturers are a major part of the program. Multibrand truck platooning in real traffic conditions and full-scale demonstrations of urban road transport automation are anticipated to follow.
- Breslin noted that many of these projects are appropriate for "twinning" with research projects in the U.S. The EC and U.S. DOT encourage twinning to exchange knowledge and experience and exploit synergies. In this approach, projects are selected and funded separately by the EC and the U.S. DOT, but include collaborative activities. Examples of twinning activities include exchanging information, data, visits, methodologies, and results. Joint workshops and publications may also be appropriate. Twinning activities are outlined in the proposals. The EU and U.S. program managers establish which projects would be of mutual interest for twinning.
- On the international level, there is a Tri-Lateral Working Group on Automation that includes Japan, the United States, and the EU. The Working Group, which is meeting at the symposium, focuses on exchanging information on experiences, plans, pilots, and other topics. There is also a subgroup addressing impact assessments.

LATEST DEVELOPMENTS IN THE STRATEGIC INNOVATION PROMOTION PROGRAM: INNOVATION OF AUTOMATED DRIVING FOR UNIVERSAL SERVICES AND RELATED ACTIVITIES IN JAPAN

Hajime Amano, *ITS Japan*

Hajime Amano provided an update on the Japanese Cross-Ministerial Strategic Innovation Promotion Program Innovation of Automated Driving for Universal Services (SIP-adus). He also highlighted other AV and ITS activities in Japan. Amano covered the following topics in his presentation:

- AD includes the application of numerous technologies. Onboard technologies are already available in many vehicles in Japan. The SIP-adus focuses on shared platform technologies addressing security, simulation, databases, and other elements.
- Amano reviewed the hierarchical structure of a digital map layered by static, semistatic, semidynamic, and dynamic timeframes. He described the framework for a dynamic map including data collection, compilation as a dynamic map, and service operations. He noted that the Dynamic Map Planning Co., Ltd., was founded in June 2016 to establish technologies and a business scheme to build and maintain the dynamic map for AD and other applications. Survey and digital map providers and automotive manufacturers participated in the process.
- Amano discussed that human factors research is an important part of the overall research program in Japan. The focus at an early stage was on the transition of the role and authority of human drivers in conditional automation and partial automation. Different use cases, scenarios and interactions, and transition events were examined. He noted that the human factors research scope was expanded to include issues that arise during the interaction between AVs and drivers, surrounding road users, and society in general. Potential issues have been identified within each of these categories for additional research.
- Amano noted that the connected and automated systems are seen as a developing process. Built-in assist systems are currently available in the market, including pre-crash braking, speed and distance control, and lane-keeping assistance. Some cooperative vehicle-to-infrastructure (V2I) and V2V assist systems are available and others are in development. These and other elements combine for advanced driving assistance and fully AD.
- Amano reviewed the development and use of traffic signal prediction systems (TSPS). He noted that use of TSPS in select locations has resulted in a 9% reduction in wait times and 10% reduction in energy consumption.
- Amano commented that right-turn collision warning systems, which alert the driver to the location of other road users, are also being developed. A pilot of the system was conducted involving 100 drivers. An assessment of the pilot reported that 78% of the drivers paid attention to the warning messages. Further, 88% of the drivers reported that the system was useful for safe driving. The analysis also identified a 54% reduction of dangerous turns.
- Amano noted that approximately 90% of highway tolls in Japan are collected electronically. The toll-collection connected services include safety assistance, real-time traffic information, and dynamic route guidance. The toll system also provides freight operator support services, including real-time location management and safe driving management. He also noted that advanced rapid transit is being pursued, including a bus rapid transit (BRT) plan for the Tokyo Bay area.

- Amano noted that guidelines for public road testing of AD systems were issued by the National Police Agency in May 2016. The guidelines outline that vehicles comply with the Safety Regulations for Road Vehicles and the Road Traffic Act. They also direct that the driver monitor the surrounding traffic and the vehicle's condition at all times, and operate the vehicle as necessary in the event of an emergency. The guidelines, which include other elements, represent a first step in addressing requirements for testing and deploying AD systems in Japan.
- Amano reported that field operations tests on public roads are scheduled to begin in 2017. Test facilities and operation management will be provided, as will dynamic map data for the test sites. Participants who test their vehicles will be required to arrange other resources. Test sites include expressways, arterials, and separate test facilities.
- Amano noted that the focus of the SIP-adus is on providing an inclusive society, where diverse people in diverse communities actively participate in generating values, enhancing both the wellness of individuals and economic development opportunities for communities. AD technologies integrated with social innovations should provide everyone with mobility to fully exercise their capacity, enabling the sustainable development of the society.
- Amano summarized that the ITS Japan Action Plan 2016–2020, which focuses on three types of areas: clusters of villages, integrated regional hubs, and megacities. The plan matches the transportation components, including ITS, to the characteristic and functions of each area.

U.S. DOT AUTOMATION AND SMART CITIES RESEARCH

Kevin Dopart, *Intelligent Transportation Systems Joint Program Office, U.S. Department of Transportation*

Kevin Dopart discussed some of the major U.S. DOT automation milestones for 2016 and highlighted examples of research projects sponsored by the U.S. DOT. He covered the following topics in his presentation.

- Dopart described the Advanced Transportation and Congestion Management Technology Deployment (ATCMTD) contained in the FAST Act. The 5-year ATCMTD Initiative makes competitive grants for model deployment sites of advanced transportation technologies to improve safety, efficiency, system performance, and infrastructure return on investment. The initiative is funded at $60 million per year. Eligible technologies include V2V, V2I, AVs, and collision avoidance systems. He noted that a local match is required and that projects must be led by a state or local government or a research consortium. The website for more information is http://www.fhwa.dot.gov/fastact/funding.cfm.
- Dopart described the Mobility on Demand (MOD) Sandbox, which is a Federal Transit Administration (FTA) program using discretionary funding. A total of $8 million was provided this year for the initiative and there are expectations that funding will be available in future years, although it is not guaranteed. The MOD Sandbox guiding principles focus on system integration, innovative business models, equity of service delivery, and partnerships. The website is https://www.transit.dot.gov/research-innovation/mobility-demand-mod-sandbox-program.html.

- Dopart noted that the Smart City Challenge, which is discussed more in the next presentation, raised the visibility for technology to address mobility, traffic congestion, and other key issues facing urban areas today. Urban automation, CVs, and intelligent sensor-based infrastructure are technology elements included in the challenge.
- Dopart reported that the U.S. DOT Intelligent Transportation Systems Joint Program Office (ITS JPO) Automation Program is focusing on the three general areas of foundational policy research, targeted technical research, and stakeholder collaboration. Examples of foundational policy research include the review of the Federal Motor Vehicle Safety Standards (FMVSS) and the development of FHWA Automated Vehicle Policy Research Needs.
- Additional policy research focuses on standards planning for automation. Examples of research in this area include identifying emerging standardization needs for AVs and developing a standards roadmap to support vehicle automation development, testing, and deployment.
- Dopart described the ITS JPO assistance with stakeholder collaboration, including the EU, United States, and Japan Tri-Lateral Working Group, the AVS 2016, and other events. The Smart City Challenge included a number of events. The agreements with the winning city of Columbus are underway. International collaboration is part of the proposal.
- Examples of targeted technical research are human factors research at FHWA and NHTSA and CACC research at FHWA. Another project developed a framework for benefits evaluation. The final Phase 1 report, Benefits Estimation Framework for Automated Vehicle Operation (FHWA-JPO-16-229), is available in the National Transportation Library (http://ntl.bts.gov/). A follow-up project includes quantitative modeling of safety benefits. Another research project is examining functional safety of automated lane-centering controls, including conventional steering, steer by wire, and conventional braking systems.
- The mixed function automation naturalistic driving study is being conducted by the Virginia Tech Transportation Institute for NHTSA. It is exploring how drivers interact with market available Level 2 automation. A total of 120 high-mileage vehicle drivers will participate for 4 weeks. Participants will receive training similar to the instruction provided at dealerships with new car purchases. In-car data will be collected to monitor driver behavior.

BEYOND TRAFFIC: THE SMART CITY CHALLENGE

Automation in the Smart City

Brian Cronin, *Federal Highway Administration*

Brian Cronin discussed the U.S. DOT Smart City Challenge. He reviewed the technology elements included in the initial applications from 78 cities, the proposals from the seven finalist cities, and the winning city's proposal. Cronin covered the following topics in his presentation:

- The Smart City Challenge was issued by the U.S. DOT on December 7, 2015. A total of 78 cities developed visions and plans to address how emerging transportation data, technologies, and applications could be integrated with existing systems in their city to address transportation challenges.

- U.S. DOT Secretary Anthony Foxx announced the seven Smart City Finalists in March 2016 at the South by Southwest event in Austin, Texas. The finalists were awarded $100,000 to support the planning and development of their final proposal. The seven finalists were Austin, Texas; Columbus, Ohio; Denver, Colorado; Kansas City, Missouri; Pittsburgh, Pennsylvania; Portland, Oregon; and San Francisco, California. Secretary Foxx announced Columbus as the winner on June 23, 2016.
- Cronin noted that one of the biggest benefits of the challenge was that it created a national dialog on cities, transportation technology, automation, and mobility. A total of 150 public- and private-sector partners were identified in the seven finalists' proposals, accounting for $500 million in potential investments. The seven proposals focused on using technology to address the problems facing the cities, including safety, congestion, mobility, housing, and access to jobs.
- Cronin described the U.S. DOT vision elements, which focused on technology, innovative approaches to urban transportation, and Smart Cities. Technology elements included urban automation, CVs, and intelligent senor-based infrastructure. Examples of innovative approaches included user-focused mobility services and choices, urban goods delivery and logistics, roadway electrification, and involved citizens. Smart land use, architecture and standards, and low-cost, efficient, secure and resilient information and communication technologies (ICT) represented Smart City elements.
- Cronin reported an analysis of the initial 78 applications indicated that 82% included vehicle automation concepts as part of their Smart City vision. Many of these applications focused on leveraging AVs to connect disadvantaged communities to jobs and other opportunities. There was also interest in infrastructure needs, connecting with other vehicles, electric AVs, and using a phased implementation approach. Issues associated with the regulatory environment were also noted in many of the applications.
- Cronin noted that all 78 proposals have been posted online. A wide range of urban automation applications were included in the proposals: podcars, shuttles, buses, and freight delivery vehicles. Given the short implementation timeframe, however, low-speed shuttles or podcars were the most common applications included in the proposals.
- Examples of transit and shared mobility applications in the initial 78 proposals included automated BRT, automated light rail, bus driver assist systems, and podcars. Freight applications focused on truck platooning on arterials, automated truck operations at ports, and other approaches addressing moving goods in urban environments.
- Proposals from the seven finalist cities included low-speed AVs supporting first-mile and last-mile connections, especially in economically challenged areas, and automated and semi-automated transit vehicles. Automated truck demonstrations and truck platooning were included in some proposals. Further, many finalists' proposed electric AVs.
- The Columbus proposal included an electric AV shuttle for the first and last mile in a new town center commercial district. The area is served by a BRT line, but a connection to the new town center is missing. Traffic signal priority for the shuttles and other enhancements are also included.
- A second element of the Columbus proposal is truck platooning on arterials with traffic signal priority. The two-truck platoons will focus on key logistic hubs. The proposal also includes multiple resources to support safety and operational testing, with The Ohio State University (OSU) taking the lead. Facilities include the Sports Pavilion and Automotive

Research Complex (SPARC), the NHTSA-owned and OSU-operated Transportation Research Center (TRC), and the OSU Center for Automotive Research website.

- Cronin noted that updated information on activities underway in Columbus and other cities will be provided on the website.

AUTOMATED VEHICLES: ACCELERATING THEIR SAFE ARRIVAL

Nathaniel Beuse, *National Highway Traffic Safety Administration*

Nathaniel Beuse discussed NHTSA research projects and related activities for developing guidance on safety and cybersecurity. He covered the following topics in his presentation:

- Beuse reviewed the NHTSA research program. He noted that human factors issues associated with assisted and piloted operating modes are being examined. Human factors concerns inside and outside of the vehicle are also being explored. Comments at the public meetings indicate that pedestrians and cyclists want to know if a nearby vehicle is operating in an automated mode. System performance issues include determining what to test and how to test it. Electronic control systems safety represents a new research area for NHTSA. The initial focus on functions has been expanded to include software. A framework for conducting benefit assessments has been developed. The testing and evaluation area includes field operational tests, the Smart City Challenge, and actual system assessments.
- Beuse noted that human factors evaluation of Level 2–Level 3 AD concepts and human factors design guidance for Level 2–Level 3 AD concepts have been completed. Ongoing activities include the naturalistic driving study, test track performance characterization tests, and a functional safety assessment of lane centering.
- Near-term AV research publications include the Target Crash Populations for Concept Automated Vehicle Functions by the Volpe Center, which estimates the residual crashes AV concepts may have the opportunity to address with respect to lower levels of automated functions with overlapping target crashes. The report on Automated Vehicle Research for Enhancing Safety provides an objective level classification methodology for concept AV functions and development of high-level safety principles by levels. The Human Factors Design Guidance for Level 2 and Level 3 AD Concepts provides guidance for developing the driver vehicle interface (DVI) for Level 2 and Level 3 concepts.
- Beuse noted that a number of projects have been completed in the area of system performance requirements research. Functional descriptions for the levels of automation, classification methods for concept automated features, and top-level principles by generic automation levels were all developed. The high-level safety principles are being expanded to develop performance requirements based on use cases. Performance-based objective tests that can evaluate near-term automated features based on their classified level are being developed. Both of these efforts are targeted for completion in 2017.
- Research addressing electronic control systems safety research includes examining design process standards and their applicability to combined function automation. Research on the functional safety of automatic lane centering systems plus ACC is also underway. Research capabilities on automotive cybersecurity are expanding. Anomaly based intrusion detection systems are being researched, as are cybersecurity of firmware updates, including over-the-air

software update mechanisms. Software design, development, implementation, validations, and learning algorithms represent another emphasis area for new research.

- Beuse noted that the NHTSA FY2017 budget request includes $50 million for vehicle electronics research and $200 million for AV pilots, which was a portion of President Obama's announced almost $4 billion, 10-year investment proposal.

TRANSPORTATION AS A SYSTEM

Reuben Sarkar, *U.S. Department of Energy*

Reuben Sarkar discussed the Transportation-as-a-System (TAAS) initiative at the DOE focusing on CVs and AVs in the broader context of transportation as a mobility system. He described the DOE's TAAS-targeted research portfolio and highlighted projects and pilots currently underway. Sarkar covered the following topics in his presentation.

- Sarkar discussed the changes impacting the transportation system. These changes make planning for the future challenging. He noted that the DOE is hosting sessions in Washington, D.C.; Berkeley, California; and Ann Arbor, Michigan, to discuss possible future scenarios and how to make choices now on technology and policy to achieve desired outcomes.
- Sarkar noted that historically the DOE has focused on vehicle-level efficiency. He commented that this focus has considered technology for the maximum efficiency of the independent unconnected vehicles, which are ultimately subject to the behaviors and decisions of drivers. He noted that the DOE has recently taken a broader system-level focus, acknowledging a future of connected and automated systems across modes. He suggested focusing on the nexus of safety, energy, mobility, and accessibility.
- Sarkar noted that the DOE is interested in the energy implications of connectivity and automation. He described preliminary research conducted by the DOE National Laboratories, which indicated a possible 90% reduction in the 2050 baseline energy consumption from CV, AV, and other technologies. He noted, however, that the research also indicated a potential 200% increase in 2050 energy consumption depending upon how CVs and AVs affect travel behavior, VMT, and other factors. He suggested that this vast range of differing energy implications indicate that more research is needed on this topic.
- Sarkar discussed the increasing complex decision environment, which includes connected travelers' vehicles, fueling and charging infrastructure, and cities and regions. He suggested that transforming complexity into clarity for decision makers is challenging in this environment.
- Sarkar discussed the five TAAS research pillars, which include CAVs, urban science, mobility decision science, vehicles and infrastructure, and multimodal. Related to the connectivity and automation pillar, Sarkar noted that the DOE is examining the energy impacts of CVs and AVs. Data on energy use is being gathered from a fleet of 500 vehicles in Ann Arbor. The data are being used in simulation and modeling for complex systems, and to inform future policy and research. Other activities focus on identifying CAV-enabled opportunities and addressing barriers to CAV deployment.

- Sarkar noted that the urban mobility science pillar focuses on developing and applying city-scale computational mobility models that will support decision makers in deploying technologies and innovative approaches.
- Sarkar described the mobility decision science pillar, noting the importance of considering human behavior and individual decision making in the development of new transportation technologies and mobility options. Research in this pillar will examine how stakeholders—consumers and others along the transportation value chain—interact with the system and how they make decisions on vehicle purchases, technology adoption, and trip and mode planning.
- The vehicles and infrastructure pillar includes using available tools and resources to identify locations for vehicle fueling and charging infrastructure. He suggested that CVs and AVs may change the existing paradigm, reducing the number of needed charging stations. Therefore, wireless charging and dynamic charging may be appropriate long-term considerations with AVs, CVs, and different ridesharing and car-sharing models.
- Sarkar noted that historically the DOE has focused on individual modes, primarily those that have the largest energy impact. The DOE has not typically considered the end-to-end trip efficiencies and the use of different modes for personal transportation and goods movement. He reported that the multimodal research pillar focuses on better understanding the energy impacts from potential mode shifts and multimodal trips as a result of CV, AV, car sharing, and ridesharing. The focus is on the energy impacts of efficient, seamless multimodal transportation for people and goods.
- Sarkar described the capabilities of the DOE National Laboratories related to simulation and modeling tools that can be used in research in the five pillars. He highlighted the new Systems and Modeling for Accelerated Research in Transportation (SMART) consortia. He noted that SMART is initially bringing together five national laboratories to explore the nexus of energy and future mobility paradigm. SMART is participating with the U.S. DOT in providing these tools to the Smart City Challenge.

DRIVING TO THE FUTURE

Karl Simon, *U.S. Environmental Protection Agency*

Karl Simon discussed the potential impacts of AVs on energy and the environment. He covered the following topics in his presentation:

- While noting the impressive safety benefits of AVs, Simon discussed the potential energy and environmental benefits from AVs. He described the challenges of meeting the Paris goal of reducing greenhouse gas (GHG) emissions by 80% below the 2005 level by 2050. He noted that while new vehicle fuel economy is at an all-time high, incremental improvements will not be enough to meet the 2050 goal.
- Simon suggested that CAVs could be part of the solution in meeting the 2050 goal. He noted that the impact of CAVs on reducing or increasing emissions is not clear. He reviewed five areas that will likely influence the impact CAVs will have on the environment. The areas are vehicle efficiency, travel demand, shared mobility, built environment, and fuels and electrification.

- Simon noted that such data indicates that AVs can operate more efficiently than human drivers, but that more research is needed to better quantify the potential energy impacts. He suggested that electrification needs to be a key part of the AV deployment to meet the 2050 goal. The transition and timing to more electric vehicles is also not well known. The impact of AVs on the built environment is also uncertain. For example, parking garages may no longer be needed and could be repurposed for other uses. More research is also needed on the opportunities and impacts of shared mobility and travel demand.
- Simon reviewed the DOE analysis on the potential impacts of HAVs on travel, energy, and carbon. The analysis indicated that factors such as vehicle right-sizing, platooning, and changed mobility services would reduce energy consumption, while reductions in travel cost would result in more travel and increases in energy consumption. He suggested that focusing on the positive aspects of AVs now would assist in ensuring positive environmental benefits as they are introduced into the marketplace. He noted that changes are occurring in the automotive industry, the energy sector, and shared services. He suggested efficient, electric CAVs could reduce emissions from the transportation sector. He further suggested that policies at the federal, state, and local levels could assist in ensuring energy and environmental benefits from AVs.

CLOSING SESSION

Reports from the Breakout Sessions

JANE LAPPIN
Toyota Research Institute, presider

The closing session featured brief summaries from each of the 22 breakout sessions listed in Table 1. The presentations followed a common format highlighting the breakout session agenda, the key takeaways and lessons learned from the presentations and discussions, and suggestions for moving forward. The presentations, which are summarized in this section, were clustered into the six topic areas highlighted in Table 2.

TABLE 1 List of Breakout Sessions

Breakout Session	Appendix with Detailed Summary
Breakout 1: Public Transport and Shared Mobility	Appendix A
Breakout 2: Law and Policy as Infrastructure	Appendix B
Breakout 3: Human Factors in Road Vehicle Automation	Appendix C
Breakout 4: Impact Assessment	Appendix D
Breakout 5: Enabling Technologies	Appendix E
Breakout 6: Safety Assurance	Appendix F
Breakout 7: Future Challenges for Automated Trucks	Appendix G
Breakout 8: Traffic Signal Control with Connected and Automated Vehicles	Appendix H
Breakout 9: Methods for Assessing the Market Acceptance, Adoption, and Usage of Automated Vehicles	Appendix I
Breakout 10: Ethical and Social Implications of Automated Vehicles	Appendix J
Breakout 11: Early Implementation Alternatives for Automated Vehicles: An Interactive Scenario-Planning Session	Appendix K
Breakout 12: AV-Ready Cities or City-Ready AVs	Appendix L
Breakout 13: Design and Operational Challenges and Opportunities for Deploying Automated Vehicles on Freeways and Managed Lanes	Appendix M
Breakout 14: Reducing Conflicts Between Vulnerable Road Users and Automated Vehicles	Appendix N
Breakout 15: Behavioral Experiments for Modeling Adoption and Use of Autonomous Vehicles	Appendix O
Breakout 16: Aftermarket Systems (ADAS-Related)	Appendix P
Breakout 17: Policy Making for Automated Vehicles: A Proactive Approach for Government	Appendix Q
Breakout 18: Effects of Vehicle Automation on Energy and Carbon Intensity	Appendix R
Breakout 19: Cybersecurity and Resilience Challenges and Opportunities for Self-Driving Vehicles	Appendix S
Breakout 20: Infrastructure, Work Zones, and Digital Infrastructure	Appendix T
Breakout 21: Traffic Flow of Connected–Automated Vehicles	Appendix U
Breakout 22: Transformational Technologies—Can Our Research Processes Keep Up?	Appendix V

TABLE 2 Breakout Group Summaries by Cluster

Cluster	Breakout Group
User-Related AV Issues	Breakout 3: Human Factors in Road Vehicle Automation Breakout 14: Reducing Conflict Between Vulnerable Road Users and Automated Vehicles Breakout 9: Methods for Assessing Market Acceptance, Adoption, and Usage of Automated Vehicles Breakout 15: Behavioral Experiments for Modeling Adoption and Use of Autonomous Vehicles
Specific AV Areas	Breakout 1: Public Transport and Shared Mobility Breakout 7: Future Challenges for Automated Trucks Breakout 16: Aftermarket Systems (ADAS-Related)
Policy and Societal Issues	Breakout 2: Law and Policy as Infrastructure Breakout 10: Ethical and Social Implications of Automated Vehicles Breakout 17: Policy Making for Automated Vehicles: A Proactive Approach for Government
Planning for AVs	Breakout 4: Impact Assessment Breakout 18: Effects of Vehicle Automation on Energy and Carbon Intensity Breakout 11: Early Implementation Alternatives for Automated Vehicles: An Interactive Scenario Planning Session Breakout 12: AV-Ready Cities or City-Ready AVs? Breakout 22: Can Our Research Processes Keep Up in an Age of AVs and Other Transformational Technologies?
Technology Issues	Breakout 5: Enabling Technologies Breakout 6: Safety Assurance Breakout 19: Cyber Security and Resilience Challenges and Opportunities for Self-Driving Vehicles
Operational Issues for AVs	Breakout 13: Design and Operational Challenges and Opportunities for Deploying Automated Vehicles on Freeways and Managed Lanes Breakout 20: Physical Infrastructure, Work Zones, and Digital Infrastructure Breakout 21: Traffic Flow of Connected Automated Vehicles Breakout 8: Traffic Signal Control with Traffic Signal Control with Connected and Automated Vehicles

USER-RELATED AUTOMATED VEHICLE ISSUES

Breakout Session 3: Human Factors in Road Vehicle Automation

Daniel McGehee, *University of Iowa*
David Yang, *Federal Highway Administration*

Summary of the Meeting Agenda

This breakout group sponsored two sessions. The first session was Look Ma', No Hands! MyCarDoesWhat? Preparing the American Public for AVs and the Changing Transportation Network. The session began with presentations from the following speakers:

- Advanced Technology Integration: A Bird's Eye View of Where We've Been and Some Observations Toward the Future. Linda Angell, Touchstone Evaluations, Inc.; Bernard Soriano, California Department of Motor Vehicles; Daniel McGehee, Ashley McDonald, Alex Epstein, University of Iowa. MyCarDoesWhat.org.
- A Large-Scale Virtual Proving Ground for Automated and Connected Vehicle Research. Omar Ahmad, National Advanced Driving Simulator, University of Iowa.
- Public Education on Truck Platooning. Oliver Bayley, Peloton Technology, Inc.

The second session was Human Factors Issues Associated with Connected Automation Applications. This session featured the following elements:

- A panel on Human Factors Findings from a Connected Automation Application–CACC:
  - Opportunities for Connected Automation Applications to Improve Mobility and Energy Use. Robert Ferlis, Federal Highway Administration.
  - Findings from CACC Experiments: Overview and Merging. Brian Philips, Federal Highway Administration.
  - Findings from CACC Experiments: Car Following and Collision Avoidance. Vaughan Inman, Leidos.
- Small group discussions and reports on discussions.
- Facilitated Expert Panel Discussion. Facilitator: Don Fisher, Volpe National Transportation Systems Center. Expert panel members:
  - Jessica Cicchino, Insurance Institute for Highway Safety;
  - Charles Green, General Motors;
  - John Lee, University of Wisconsin;
  - Michael Regan, ARRB Group;
  - James Sayer, University of Michigan Transportation Research Institute; and
  - Heishiro Toyoda, Toyota Technical Center.

Summary of Key Takeaways and Lessons Learned from the Breakout Discussion

- Participants in the first session discussed the need for public education and outreach to support the introduction and use of automated, connected, and autonomous vehicles. It was suggested that many drivers do not understand current technologies available on their vehicles. Providing information on both the technology that is available on a vehicle and how it works and on technologies on other vehicles, such as truck platooning, and on the infrastructure, was noted as important by participants.

- Participants in the first session also discussed the need to integrate consumers into the design and development process for automated, connected, and autonomous vehicles. The need to bring together OEMs, software developers, government agencies, academia, and marketing was suggested as important by participants.
- Participants in the second breakout group discussed the major human factors issues that need to be addressed with the design, development, and deployment of connected automation applications.

Moving Forward

- Participants discussed the need to continue human factors research associated with a range of topics related to connected, automated, and autonomous vehicles.

Breakout Session 14: Reducing Conflict Between Vulnerable Road Users and Automated Vehicles

Ryan Greene-Roesel, *San Francisco County Transportation Authority*

Summary of the Meeting Agenda

This breakout session featured the following four presentations followed by a discussion of issues and opportunities:

- What is the current state of pedestrian–vehicle interaction? Justin Owens, Center for Vulnerable Road User Safety, Virginia Tech Transportation Institute.
- Vulnerable Road Users in the Age of Automated Vehicles: How to Ensure Safe Interactions? Azra Habibovic, Viktoria Swedish Information Communication Technology Institute.
- Connected Pedestrians at Signalized Intersections in a CAV Environment. Larry Head, Systems and Industrial Engineering, University of Arizona.
- Prospect: PROactive Safety for Pedestrians and Cyclists: Project Overview. Andres Apracio, Product Manager, Applus IDIADA.

Summary of Key Takeaways and Lessons Learned from the Breakout Discussion

- Participants discussed that some groups have suggested that traffic signals will not be needed in the future with AVs, but it is unclear how those intersections would safely accommodate VRUs.
- Participants discussed that OEMs and researchers are developing different approaches for communicating a vehicle's intent to VRUs and other road users. Participants noted that it was not clear if work is being done or should be done to develop methods for VRUs to clearly signal their intent to a vehicle.

- Participants noted that significant progress is being made in testing and implementing pedestrian crash avoidance systems.

Moving Forward

- Participants discussed the need to develop a vision for the multimodal intersection of the future that safely accommodates CAVs and VRUs.
- Participants suggested the need for research help develop a universal language for CAV and VRU communication.
- Participants discussed the need to pursue safety improvements now, without waiting for future CAVs.

Breakout Session 9: Methods for Assessing Market Acceptance, Adoption, and Usage of Automated Vehicles

Johanna Zmud, *Texas A&M Transportation Institute (TTI)*

Summary of the Meeting Agenda

This breakout session focused on the challenge of collecting reliable and accurate information on AV market acceptance and adoption for use in travel forecasting models when people generally have little knowledge of and no experience with AVs. The following three presentations provided state-of-the art summaries of measuring market acceptance, adoption, and usage of AVs. All three speakers described the use of multimethod modeling scenarios, quantitative surveys, and qualitative methods.

- Revolution in the Driver's Seat. Mike Quinn, BCG. This presentation examined the modeling of 2035 global market penetration based on survey and expert interviews, together with scenarios to account for influencers on penetration.
- Exploring User Expectations of Autonomous Driving. Stefan Trommer, DLR. This presentation described the use of focus groups to develop scenarios to derive inputs to vehicle fleet and travel demand models in the United States, Germany, and China.
- Forecasting Consumer Adoption of Automated Vehicles and the Impact on Total Vehicle Sales. Dan Weinstrin, RSG. This presentation examined the use of the joint revealed–stated-preference modeling approach to test sensitivity to owned AV and pay-per-use attributes and to model adoption and sales impact.

The survey results of a pre-symposium survey of the breakout group leaders on priority research questions for assessing vehicle ownership and usage were reviewed. Small facilitated discussion groups were used to identify possible research opportunities, with reports given back to the full group.

Summary of Key Takeaways and Lessons Learned from the Breakout Discussion

- Participants discussed that although the methods presented represent the state-of-the-art, they are not adequate to really understand the questions associated with market acceptance, adoption, and usage of AVs. Vehicle use and vehicle ownership were identified as some of the key topics for further research.

- Participants discussed that AV usage questions are very important. Participants suggested that key unknowns include public acceptance, incentives, and disincentives to the use of both private and shared AVs, the value of time, and examining use across purposes and across modes. All of these topics were noted as needing further research.
- Participants stressed the need for continuous studies of AV use over time to determine trends and potential dynamics in the marketplace. Leveraging any pilot tests to carefully study user behavior was stressed as important by participants.
- Participants also stressed the need for creativity in developing new methods to anticipate and capture behavior changes, including new trips, travel purposes, and influences on lifestyles. One suggestion made by participants was to use creative methods find behavior analogs, such as vouchers for chauffeurs, to study potential AV acceptance.
- The need to educate respondents prior to actual data collection was identified as important by participants. The public's mental models of AVs are very different from how AVs might actually operate. It was suggested that existing surveys are collecting information on people's current perceptions of AVs, with little understanding of how these vehicles might actually operate.
- Participants discussed the need to understand AV use first in order to understand AV ownership. Key unknowns related to AV ownership discussed by participants included willingness to pay, ownership persistence, and the size and impact of new owner groups such as the mobility impaired and older and younger individuals. Participants noted that research on these topics would be beneficial.

Moving Forward

- Participants discussed the need to continue the dialogue on this topic. It was suggested that a breakout session at the 2017 AVS focusing on hypothetical user decisions and creatively developing methods to obtain that information would be beneficial.
- Participants identified the need to collaborate on new research on the topic focusing on mixed method, cross disciplinary, and cross-national approaches. Exploring diverse research funding sources and communicating the importance of creative problem solving to sponsors and decision makers was noted as important.
- Tracking, compiling, and communicating examples of behavior analogs, creative methods, and other related approaches was also identified as important by participants.

Breakout Session 15: Behavioral Experiments for Modeling Adoption and Use of Automated Vehicles

Yoram Shiftan, *Technion University, Israel*

Summary of the Meeting Agenda

This breakout session focused on bringing together travel demand and behavioral researchers to discuss behavioral experiments to support modeling, planning, and policy for AVs. The session included speakers presenting different approaches including simulation-based scenario analysis studies, stated-preference surveys, virtual reality and game simulators, revealed-preference surveys, and naturalistic experiments. Participants discussed possible impacts of AVs on vehicle ownership and use, travel behavior, mode of travel, activities and lifestyles, and land use. Topics for further research were identified by participants.

Summary of Key Takeaways and Lessons Learned from Breakout Discussion

- Participants discussed that individual behavior is key to the impact AVs will have on the transportation system, land use, and human health. Depending on individual behavior, the benefits may be positive, negative, or mixed.
- Participants discussed the advantages and limitations of different research methods. It was suggested that multiple approaches are needed to better understand the potential impacts of AVs on traffic congestion, mobility, land use, and the environment.

Moving Forward

- Participants suggested that an integrated approach using multiple methods would provide the most robust data to address key questions. It was further suggested that additional research was needed to outline the key elements of this integrated approach.
- Participants identified the need to develop better methods to provide experience and knowledge of AVs to respondents. It was suggested that research developing these methods would be beneficial.
- Participants discussed that an individual's knowledge, awareness, and preferences will change over time. As a result, participants identified the need for research and experiments to continue over time and to collect consistent longitudinal data and data across geographies.
- Participants discussed the need for coordination and collaboration with other research activities, including leveraging field operational tests for behavioral research. It was suggested that including research in travel, activity, attitude, and behavioral changes in all field operational tests and pilots would be beneficial.
- Participants discussed the need for research to develop a standard set of before-and-after questions to use in all experiments.

AUTOMATED VEHICLE APPLICATION AREAS

Breakout Session 1: Public Transport and Shared Mobility

Gary Hsueh, *Arup*

Summary of the Meeting Agenda

This breakout group sponsored two sessions. The first session included eight presentations addressing updates on research, projects, pilot programs, and testing sites. Information on program updates and funding opportunities was also presented. Speakers included Susan Shaheen, University of California Berkeley; Randy Iwasaki, Contra Costa Transportation Authority; Art Guzzetti, American Public Transit Association; Justin Holmes, Zipcar; Robert

Lohmann, 2getthere; Adriano Alessandrini, University of Rome; Chris Kopp, HNTB; Chris Augenstein, Virginia Transportation Authority; Gary Hsueh, Arup; Vincent Valdes, FTA; Michael McGurrin, Noblis; and Doug Gettman, Kimley-Horn. The first session also featured a "Shark Tank" on what it takes to launch a successful AV pilot program, with several individuals making pitches on their approaches.

The second session included two panels. The first panel was on the integration of public and private models. It included Emily Castor, Lyft; Susan Shaheen, University of California, Berkeley; Barbara Laurenson, Metropolitan Transportation Commission; Michale Scrudato, Munich Re, and Matthew George, Bridj. The second panel focused on public transport in the future. Speakers included Stan Young, National Renewable Energy Laboratory (NREL); Sam Lott, Texas Southern University; Jerome Lutin, New Jersey Transit (retired); Mark Mindorff, Dallas Area Rapid Transit; and John Mirisch, Mayor of Beverly Hills, California. The session also included a workshop on policy implications and possible research for public transport and shared mobility.

Summary of Key Takeaways and Lessons Learned from the Breakout Discussion

- Participants discussed that shared AVs will begin to blur the lines between public and private transportation. This approach could deliver efficient and affordable public transportation to meet societal needs—improving access to jobs and healthcare. Participants noted that it would not replace mainline public transportation, however.
- Participants discussed deployment opportunities for first- and last-mile connections, as well as addressing the needs of underserved populations and areas lacking quality transit service. Participants noted that these opportunities serve a much broader market.
- Participants suggested that cities and sites are different, so that shared AV deployments need to be tailored to varying technical, cultural, and legal contexts.
- The strong interest at the local level in testing and deploying shared AVs was noted by many participants.
- Participants discussed that pilot programs, enabled by public–private partnerships, are encouraging private shared services to adapt and expand their functionality to meet the needs of public transit users.
- It was noted by participants that there are many competitive federal funding opportunities that highlight the role of shared AVs in public transportation.
- Participants suggested that significant thought is required in setting and adapting policies at all levels of government to support the use of automation and shared AVs in public transportation.

Moving Forward

- Developing and adopting a traveler-centric approach to transportation planning that stresses equity, occupancy, and sustainability was suggested by participants as future research.
- Participants discussed the need to promote public–private partnerships that leverage the strengths unique to each sector.
- Participants noted that it would be beneficial to add flexibility in transport procurement processes to consider mobility as a whole.

- Participants suggested that developing safety standards to enable the implementation of automated technology in public fleets was needed.
- Breakout group participants encouraged the use of pilot programs as a safe space for experimentation.
- Participants discussed the importance of measuring, documenting, and sharing best practices and impacts of automated shared mobility and public policy adaptations.

Breakout Session 7: Future Challenges for Automated Trucks

Richard Bishop, *Bishop Consulting*

Summary of the Meeting Agenda

This session focused on challenges and opportunities for deploying various levels of automated trucks. The following four general topics were discussed:

- The EU Truck Platooning Challenge was discussed in more detail following the presentation in the general session.
- Recent studies from the Organisation for Economic Co-operation and Development (OECD) on truck automation policy options and labor market effects were presented and discussed.
- A panel discussed truck cooperative automation issues and opportunities.
- A discussion on full automation and architecture aspects provided more focus on engineering issues.

Participants in the breakout session include shippers, truck OEMs, technology developers, federal agencies, states DOTs, and analysts and researchers.

Summary of Key Takeaways and Lessons Learned from the Breakout Discussion

- A key question from the discussion of the EU Truck Platooning Challenge was if it would be possible to undertake a similar event in the United States? Participants discussed possible benefits for such events and the potential outcomes, which might include policy, deployment, and operational guides.
- Participants suggested that examining traffic interactions via "learning by doing" was important. The Dutch example of this approach was noted as one possible model for use in the United States.
- Panelists discussing cooperative truck automation noted that V2V was key for platooning and that V2I plays a key role in supporting urban operations such as freight signal priority.
- Participants discussed that standards are key for truck platooning between brands and between fleets.
- Participants discussed independent truck automation and suggested that truck drayage operations at intermodal centers could be the first deployments of this approach. It was noted by some participants that the Florida DOT is examining this application.

- Participants discussed more advanced levels of automation and truck platooning. It was noted that the military is focusing on driverless vehicles, whereas shippers view this approach as a long-term option.
- Participants discussed that verification and validation were hard, and that collaboration across industry was important to advance verification and validation.

Moving Forward

- Participants discussed opportunities for collaboration between the private sector and the military, as well as between commercial trucks and passenger vehicles.
- Participants identified a need to examine HMI aspects of the evolving driver role with truck platooning.
- Exploring near-term, high-automation truck platooning deployments in limited geographic settings was also noted as important by participants.
- Participants suggested that continuing to focus on the interoperability of cooperative systems in the long term was also important.

Breakout Session 16: Aftermarket Systems (ADAS-Related)

Jim Misener, *Qualcomm Technologies Incorporated*

Summary of the Meeting Agenda

This session focused on obtaining a better understanding of the role that aftermarket systems may play in accelerating the deployment of AVs. The session brought together speakers from Silicon Valley's start-up companies, established Tier 1 suppliers, and commercial fleet operators to discuss the benefits and challenges associated with aftermarket system deployment. Consideration was also given to business models, since aftermarket systems cannot be expected to be mandated by government.

The session featured the following seven presentations and a panel composed of the following speakers:

- Return of the Aftermarket. Roger Lanctot, Strategy Analytics.
- Aftermarket Truck Platooning and Automation. Steve Boyd, Peloton Technology, Inc.
- The Utility, Economic, and Business Models for Aftermarket ADAS Today and in the Near Future. Chris Brogan, AssureNet.
- Driving Behavior Detection Technology for ADAS/AD and its Application for Smarter Mobility. Hiro Onoda, SVP Denso International America, Inc.
- The Future of Urban Mobility is Not Automomy, It's Autonomous, Connected, Electrified, and Shared. Stefan Heck, Nauto.
- The Fleet Operator's Perspective on Retrofitting Aftermarket ADAS technologies. Scott Perry, Ryder Fleet management Solutions.
- Role of V2X-Based Aftermarket Safety Devices (ASD) in Smart Cities. Paul Sakamoto, Savari.

Summary of Key Takeaways and Lessons Learned from the Breakout Discussion

The speakers provided the following market insights during their presentations and the panel discussion:

- Lunctot described aftermarket benefits from new technologies, especially camera systems.
- Brogan described ADAS payback opportunities.
- Onoda discussed data-enabled mobility services.
- Boyd described truck platooning and related cloud services.
- Perry noted that there is a need for ADAS aftermarket systems, but that legacy equipment, operational policies, and drivers make it challenging.
- Heck described an approach using aftermarket ADAS in combination with collecting data on good and bad driving behaviors to develop a harmonious autonomous vehicle.
- Lanctot noted that China is potentially a very large market for aftermarket ADAS.
- Sakamoto suggested that DSRC is the key pathway.

DSRC aftermarket safety systems, especially on top of already existing devices consumers use, such as smartphones and personal navigation devices, were described by several speakers.

Moving Forward

- Participants suggested that developing a catalog of the diverse aftermarket opportunities would be beneficial. The catalog could be categorized by technology, such as DSRC-based and camera-based systems, and by mode, fleet, and user.
- Participants suggested that identifying the progression of each of these opportunities to different levels of automation would also be beneficial.

POLICY AND SOCIETAL ISSUES

Breakout Session 2: Law and Policy as Infrastructure

Karlyn Stanley, *RAND*

Summary of the Meeting Agenda and Key Topics Discussed

This breakout group sponsored two sessions. The following panels were featured in the sessions.

Panel 1: What Road Authorities Can Do to Prepare for Automated Vehicles

Speakers on this panel included Paul Hemmersbaugh, NHTSA; David Strickland, Venable, LCP; Tilly Chang, San Francisco Country Transportation Authority; Steve Buckley, City of Toronto; and Sarah Hunter, X. Bryant Walker Smith, University of South Carolina, moderated the panel. Hemmersbaugh suggested that state, local, and municipal authorities should focus on areas of traditional regulation. The panelist noted that while there are some gray areas, most regulations are clear. Hunter advised road authorities not to panic, as most laws are fit for purpose. Strickland suggested that it was important for states and road authorities to collect data and that data creation could be helpful for the future of AV deployment and vehicle safety. Chang advised municipal, local, and state authorities to know their policy framework, and goals and to stick with them. The panelists suggested that authorities should consider how licensing, traffic laws, safety inspections, and other traditional functions might be performed differently as vehicles with varying degrees of automation become part of the roadway scene. How to license a driver was one of the key issues discussed by panelists. Defining the state and the federal spheres of authority for regulation and the safe operation and testing of AVs was another topic discussed.

Panel 2: What, Not Who, Are You Licensing as a Driver?

This panel explored the concept of the driver from a U.S., Australian, EU, and Japanese perspective. The speakers included Paul Hemmersbaugh, NHTSA; Tom Al Kim, RDW, Netherlands; Marcus Burke, Nautical Transport Commission, Australia; Maxime Flament, ERTICO, Belgium; and Shin Morishita, Cabinet Office of the Japanese Government. James Anderson, RAND, moderated the panel. The key issues discussed by all the panelists were responsibility and control of the vehicle. A recent discussion paper on automated vehicles with and without a driver was described. Information on the EU's approach and work developed in Australia and in Japan was also presented.

Panel 3: Balancing Security, Privacy, and Innovation in Automated Vehicle Data Use

Speakers on this panel included Erinn DePorre, Chrysler Fiat; Jonathan Petit, Security Innovation; Dorthy Glancy, Santa Clara University School of Law; and Ryan Falconer, Arup, Canada. Karlyn Stanley, RAND, moderated the session. The panelists provided definitions of AV and CV data and discussed the differences in ownership and privacy associated with data from various sources. Panelists and participants discussed the interaction and the intersection of data security, privacy, and innovation. Panelists discussed whether data anonymization was a "silver bullet" or an illusion. Participants developed the key elements of a CV data management plan for a hypothetical metropolitan area.

Breakout Session 10: Ethical and Social Implications of AVs

Suzie Lee, *Virginia Tech Transportation Institute*

Summary of the Meeting Agenda

This session included four panels, with three speakers each. The panelists provided short remarks, followed by questions and discussions with all the participants. The session was the first time the topic of the ethical and social implications of AVs was included in the AVS breakout groups.

- Panel 1: Why Ethics? The panelists included Emily Frascaroli, Ford Motor Company; Stephen Wu, Silicon Valley Law Group; and Tom Powers, University of Delaware. Patrick Lin, Cal Poly, moderated this session.
- Panel 2: Values and Weights. This panel included Noah Goodall, Virginia Transportation Research Council; Stephen Erlien, Peloton Technology; and Sarah Thornton, Stanford University. Ryan Jenkins, Cal Poly, served as the session moderator.
- Panel 3: Licensing and Testing. Wendy Ju, Stanford University; Suzie Lee, Virginia Tech Transportation Institute; and Shad Laws, Renault Innovation Silicon Valley were the panelists. Keith Abney, Cal Poly, moderated the session.
- Panel 4: Consumer Perceptions. Sarah Hunter, X; Joe Barkai, IDC; and Jason Millar, University of Ottawa were the panelists. Patrick Lin, Cal Poly, moderated the session.

Summary of Key Takeaways and Lessons Learned from the Breakout Discussion

- Participants discussed that there are no easy answers to ethical dilemmas, and that no answer is black and white.
- Participants suggested that ethical decisions are not being hard coded into AV systems. It was further suggested that based on the optimization scheme (safety principles, efficiency principles, energy principles, or other principles) the actions taken by the vehicle may appear to be making a decision based on ethical considerations.
- Participants discussed the potential to leverage advances made in bioethics over the past 50 years to help address the potential issues with AVs.
- Participants suggested that OEMs and developers should employ ethicists to ensure that these discussions are occurring during the design phase, going above and beyond the minimum legal standard of care, which would help alleviate possible product liability concerns.
- Some participants suggested that all data may need to be open to public and other researchers.

Moving Forward

- Participants discussed that standards are needed if data is to be shared. The standards would address the type and amount of data, the format, and other elements.
- Participants suggested that on-road testing was needed after computer modeling and simulations. It was also noted that human subject protection considerations will need to be addressed with tests and pilots.
- Participants supported continuing the session at future symposiums. Possible topics for future meetings included lessons learned from on-road testing, lessons learned from any crashes and incidents, and new ethical considerations based on new technological advances. Participants also suggested that ongoing critical thinking was needed on the topic.

Breakout Session 17: Policy Making for Automated Vehicles—A Proactive Approach for Government

Anita Kim, *U.S. DOT Volpe Center*

Summary of the Meeting Agenda

This breakout session focused on the key challenges facing transportation agencies related to AVs and the approaches agencies can use to enable the desired transportation systems. It included an overview of the current policy landscape for AV and two panels. The first panel focused on policy perspectives and the second panel addressed policy challenges currently facing transportation agencies. The breakout session concluded with an interactive discussion on the future policy levers state and local agencies can use to incentivize or deter certain behaviors and outcomes.

Panel 1: Policy Perspectives

Panelists in this session included Robbie Diamond, Securing America's Future Energy; David Strickland, Self-Driving Coalition for Safer Streets; Paul Scullion, Association of Global Automakers, Inc.; Jill Ingrassia, AAA; and Chandra Bhat, University of Texas at Austin.

Panel 2: Real Policy Challenges from Real Agencies

Panelists included Nathaniel Beuse, NHTSA; Bernard Soriano, California Department of Motor Vehicles; Tracy Larkin-Thomason, Nevada Department of Transportation; Mike Alexander, Atlanta Regional Commission (ARC); and Karla Taylor, City of Austin Transportation Department.

Summary of Key Takeaways and Lessons Learned from the Breakout Discussion

- Participants discussed that AVs are increasingly becoming a part of the public conversation and that many states are considering legislation related to AVs.
- The importance of public and public agency education was noted as important by many participants.
- Participants discussed that the technology development and adoption paths remain unclear, making planning for the future difficult.
- Participants described possible risks associated with prematurely regulating or legislating AVs and related technologies.
- The potential that cities could become the early adopters and test beds for AVs was suggested by participants.
- The importance of incorporating AVs into the long-range transportation planning process and polices was discussed by participants.
- Participants suggested that accepting the new normal was important for agency staff, policy makers, and the public.

Moving Forward

- Participants suggested the need to develop accurate and timely information on AVs to educate and inform policy makers, agency staff, and the public.
- Reviewing existing regulations and laws was noted by participants as an important first step for agencies at all levels. Sharing this information across jurisdictions to learn from others was also suggested.
- Participants discussed that a good approach was focusing first on where agencies traditionally play a role and building on existing strengths.
- Participants noted that engaging with local communities was important.
- Experiencing the technology directly was also suggested as a good approach for agency staff and policy makers. The benefits of firsthand experience were noted by some participants.
- Participants identified the need for research on long-range planning models and processes to address AVs.
- Sharing information, data, and experiences in order to learn was noted as important by participants.

PLANNING FOR AUTOMATED VEHICLES

Breakout Session 4: Impact Assessment

Satu Innamaa, *VTT Technical Research Centre of Finland*

Summary of the Meeting Agenda

The goals for the session included applying systems thinking to AV impacts and identifying the most important linkages between direct and indirect impacts. Another goal was to improve the design of field operational tests to maximize the insights obtained, including the appropriate performance indicators and the establishment of a baseline. Based on the impacts and their outcomes, another goal was to identify some of the investment and policy decisions needed today to make good future outcomes more likely.

The session included two parts. The first part focused on direct impacts. Five AD cases were used to assess the direct impacts. These cases were automated low-speed shared shuttles, automated local delivery, truck platooning, passenger vehicle mid-level automation, and fully automated private vehicle. The second part of the session examined indirect impacts using the use cases from the first part.

Participants developed an approach to assess the direct and indirect impacts from AVs. Technology, user attitudes, and policy and investment decisions combine to support the deployment of automated scenarios and use cases, which have direct impacts that can be measured. Examples of direct impacts include costs, infrastructure, safety, vehicle operations, energy and emissions, and personal mobility. The direct impacts have indirect impacts, which might include network efficiency, travel behavior, public health, land use, and other factors. The infrastructure also influences the indirect impacts. The direct and indirect impacts lead to outcomes that result in infrastructure investments and policy decisions, which in turn influence the development of new automated use case scenarios.

Summary of Key Takeaways and Lessons Learned from the Breakout Discussion

- Participants discussed that the short duration of most field operational tests makes fully assessing possible impacts difficult.
- Participants discussed that impact mechanisms are complex and far-reaching, and include direct and indirect impacts, and short-term and long-term impacts. Assessing the long-term impacts on society was noted as challenging.
- Participants suggested that the most important impacts are different for different people. A positive impact for one person may be a negative impact for someone else.
- Clearly defining the use case scenarios and the context (such as environment, time-scale, perception, etc.) was noted as essential, as most people have not experienced them first hand.
- Participants also noted that considering future uncertainty in today's policy and infrastructure decisions was a challenge.

Moving Forward

- Participants identified research topics that would help advance AV impact assessments.
- Participants noted that there is still a need to identify the high-level indirect impact mechanisms of AVs.
- Participants agreed that the impact assessment framework developed could facilitate the sharing of insights from pilots, demonstrations, and deployments.

Breakout Session 18: Effects of Vehicle Automation on Energy and Carbon Intensity

Paul Leiby, *Oak Ridge National Laboratory*

Summary of the Meeting Agenda

The session included the following presentations and extended discussions:

- Energy and Emissions Implications of CAVs: Results from a National Center for Sustainable Transportation Study. Matthew Barth and Caroline Roden, University of California.
- Estimate of Fuel Consumption and Greenhouse Gas Emissions Impact from an Automated Mobility District. Yuche Chen, NREL.
- Shared Autonomous Electric Vehicle Operations Across the Austin, Texas Network with a Focus on Changing Infrastructure Decision. Kara Kockelman, University of Texas at Austin.
- Anticipating the Impacts of Autonomous Vehicles Using the MOVES Model. Jun Liu, University of Texas at Austin.
- Impact of Connectivity and Automation on Advanced Vehicle Energy Use. Dominik Karbowski, Argonne National Laboratory.

- An Update on Eco-Control Strategies in the Lab and in the Field. Xueqei Qi and Matthew Barth, University of California Riverside.
- Scaling of Platooning Energy Benefits with Compatible Vehicle Adoption. Dan Mackenzie, University of Washington.
- Potential Energy Impacts of Connected and Automated Vehicles: Opportunities and Approaches. Jeff Gonder, NREL.

Summary of Key Takeaways and Lessons Learned from the Breakout Discussion

- Speakers noted that there are complicated and countervailing system level effects from automation and that the magnitude and total direction of change is unclear. Speakers reviewed many of the CAV technologies that can increase operational energy efficiency. Many changes focus on reducing congestion, smoothing traffic, and reducing zero dynamic losses at higher speeds.
- Speakers discussed the energy impacts of automated mobility services. The trade-offs in system design, such as vehicle range, charging time, and vehicle availability were discussed, along with how these influence wait time, service, and energy and emissions. Shared vehicles in the context of the internal combustion engines for last-mile service, including energy savings, were noted by some speakers.
- Improving operating efficiency through automation was examined by speakers and participants, including exploring smoothed drive cycles in simulations of urban traffic models. Scaling these insights up and integrating effects from specific cases to nationwide efforts were discussed. It was noted that benefits vary, often increasing with the scale of adoption. For example, Mackenzie reported that the per-vehicle savings increase with adoption rates in scaling up truck platooning technology.

Moving Forward

- Participants suggested continuing the dialog on the possible energy and emission impacts of CAVs at future meetings and conferences.
- Participants suggested developing research priorities into problem statements for possible funding sources.

Breakout Session 11: Early Implementation Alternatives for Automated Vehicles—An Interactive Scenario Planning Session

Caroline Rodier, *University of California, Davis*

Summary of the Meeting Agenda

This session drew on qualitative scenario planning methods to explore the implementation of near-term applications of AV technologies in specific geographic contexts and time horizons. Three scenarios for future applications were presented by the moderators—complete streets, first- and last-mile transit access, and managed lanes. All of the scenarios were based in California. Participants self-selected into small groups to discuss possible AV systems to address the different scenarios. The session concluded with each group presented their suggested approaches.

Summary of Key Takeaways and Lessons Learned from the Breakout Discussion

Complete Street Scenario The complete street scenario focused on downtown Los Angeles, California, in 2040. Due to the previous implementation of a complete streets program and a bike-sharing program, as well as the rapid penetration of personal autonomous vehicles, bicyclists, pedestrians, and vehicles all had equal access to local streets. This situation resulted in downtown streets approaching gridlock. To address this concern, the city recently limited vehicle access in the urban core to a fleet of small electric autonomous shared-ride or shared-taxi vehicles.

Participants discussing this scenario proposed demand responsive shared AVs and continually operating shuttles in the urban core. They also suggested a phased-in introduction to build public and business support. Other topics discussed included allowing disabled individuals to use AVs at no cost, implementing high-quality transit service to the downtown core, addressing a revenue loss to the city from the removal of on-street parking, and ensuring the safe interaction of the autonomous vehicles, bicyclists, and pedestrians.

First- and Last-Mile Access This scenario focused on the South Coast region in 2025. An efficient rail transit network is in operation, providing high-quality service linking communities in the region. While development has occurred around most stations, low-density suburbs are the norm for most of the region. Park-and-ride lots are full and limited first- and last-mile transit service has had mixed success. Low-speed AVs are being considered to provide timely and affordable first- and last-mile service.

Participants in this scenario group developed a service concept with autonomous shuttles sharing dedicated lanes with bicyclists in a hub-and-spoke system around rail stations. Autonomous shared-ride vehicles and ebike shared-use service would also be available and have access to the dedicated lanes. Participants suggested that the current parking lots at stations would be redeveloped for housing and businesses. Participants also recommended that the shuttles would provide delivery service for local businesses during off-peak time periods. Possible funding mechanisms for the service were identified by participants.

Managed Lanes The managed lanes scenario focused on Southern California in 2020. An extensive network of managed lanes exists in the region. Although viable, market penetration of AVs has been slow. The concept of AV road trains using electronic tolling is being considered to incentivize the adoption of AVs.

Participants in this scenario group identified a number of goals for the autonomous road trains in managed lanes project focusing on increasing the adoption of AVs, improving the environment, ensuring equity, improving economic efficiency, and improving safety. Concerned about public acceptance and political feasibility, participants suggested using a pilot or demonstration approach focusing on the long-distance commute market. Possible implementation issues discussed by participants included mixed-use operation, establishing and disbanding platoons, credentialing, data sharing agreements, security, and cost.

Moving Forward

- Participants suggested that additional multidisciplinary research examining additional scenarios would be beneficial. The research of these scenarios could be developed into pilots and demonstrations of different approaches to advancing automated transportation.

Breakout Session 12: Automated Vehicle–Ready Cities or City-Ready Automated Vehicles?

Siegfried Rupprecht, *Rupprecht Consulting, Germany*

Summary of the Meeting Agenda

This breakout session examined how cities can create an environment to ensure that CAVs deliver the anticipated safety and mobility benefits. The session was divided into the following four sections:

- Representatives from the International Transport Forum, Transport for London; San Francisco, California; Columbus, Ohio; and the U.S. DOT discussed city needs and city experiences.
- Scenarios of AV introduction were discussed by a representative from Toronto, Ontario, Canada.
- Panelists from Lyft and Qualcomm described the role of the private sector and industry.
- A facilitated discussion focused on action planning and future research.

Summary of Key Takeaways and Lessons Learned from the Breakout Discussion

- Participants discussed that automation is getting on the urban policy agenda. They questioned if city policies are on the automation agenda of industry, academia, and government.
- Participants discussed that automation is also a tool to implement housing, land use, and health policies, as well as addressing other critical needs in cities. Participants further discussed which transport problems automation can help to solve more effectively than other technologies or policies, such as clean fuel fleets and mobility as a service.
- The possible turning point where AVs are generating an impact on urban form was explored by participants.
- Participants discussed potential challenges with automation in city environments. It was noted that avoiding unintended impacts, while dealing with uncertainties was one challenge. The need for planning tools and assumptions represented another challenge, as did the need to make long-term infrastructure decisions now while keeping pace with many technology advancements.
- Participants discussed that the coexistence of AVs and conventional vehicles will likely be a long, and potentially messy, period.
- Participants suggested that in order to benefit from automation, cities must proactively promote behavior change and create an enabling policy framework.

Moving Forward

- Participants outlined a potential urban automation agenda. Elements of this agenda included developing a vision of how automation can help transform mobility in cities, and how to promote behavior changes through shared mobility services, quality transit development, and demand management policies. Participants suggested that supporting local shared-vehicle pilots, including identifying impacts and upscaling when successful should be included in the agenda. Participants suggested that one approach would be to consider AV-enabled corridors using supporting policies and technologies and discussed how these corridors could develop over time and interact with other modes. Participants also noted that developing scenarios for reusing parking spaces and how transit infrastructure might evolve would be beneficial.
- Participants discussed that cities will need to consider obsolescence, resilience, and new liability issues related to an AV-ready infrastructure.
- Participants suggested that the automation of freight and transit be given special attention, including considering new service models and regulations.
- Participants discussed the importance of addressing public perceptions and involving stakeholders at all levels in developing AV-ready cities.
- Participants discussed research to develop a city-focused automation work plan with partners and stakeholders.
- Participants suggested that continuing to engage in joint learning and in dialogue with industry, academia, and government would be beneficial to all groups.

Breakout Session 22: Can Our Research Processes Keep Up in an Age of Automated Vehicles and Other Transformational Technologies?

Tammy Trimble, *Virginia Tech Transportation Institute*

Summary of the Meeting Agenda

The session began with the organizers reviewing previous activities. The session was part of an ongoing effort to identify steps that agencies, TRB, and other groups can take to help enhance the ability of transportation agencies to conduct and deliver quality research within timeframes consistent with the pace of rapidly evolving transformational technologies, especially AV technologies.

Participants discussed the areas of the greatest potential to move to a timelier and strategic research approach, the possible ways to accomplish that transformation, and potential models from other sectors that should be considered.

Summary of Key Takeaways and Lessons Learned from the Breakout Discussion

Participants identified the following potential approaches to improve the research process:

- The EU Public–Private–Academic Partnership Model was suggested by participants as one approach to speed up the existing processes and to enhance the effectiveness.
- Participants suggested that it was important to recognize the different expectations of government agencies and the private sector on research projects.

- Participants discussed the need to provide clearer definitions of the strata of research that would accommodate larger and more complex projects, as well as smaller projects with shorter time frames.
- Participants also discussed recognizing the need for a different research culture that is more nimble and able to adapt to transformational technologies research projects.
- Participants discussed making the research needs process more continuous and more visible. A research needs dashboard was suggested as one approach.

Moving Forward

- The comments and suggestions from the session will be synthesized and presented to the TRB Conduct of Research Committee, which is coordinating the overall effort.
- The committees will use additional methods to share the results, which might include a summary flyer, a more in-depth report, and presenting the summary at meetings, conferences, and webinars.

TECHNOLOGY ISSUES

Breakout Session 5: Enabling Technologies

Jim Misener, *Qualcomm Technologies, Inc.*

Summary of the Meeting Agenda

One goal of this session was to explore a wide range of technologies needed to establish AVs including position, localization and mapping; algorithms, deep learning techniques, sensor fusion, guidance, and control; hybrid communications; sensing and perception; and data ownership and privacy. A second goal was to gain an understanding of how these technologies will work together to address the needs of the various applications, with recognition of data ownership, regulatory, and standardization perspectives. A third goal was to realize the potential shortfalls in these technologies.

The following topics and speakers were included in the session:

- Position, Localization, and Mapping. Mark Tabb, HERE, Sravan Puttagunt, Civil Maps, and Rob Hranac, Swift Navigation.
- Algorithms, Deep Learning Techniques, Sensor Fusion, Guidance, and Control. Trevor Darrell, PATH, Serafin Diaz, Qualcomm Technologies, Inc., and Shahrokh Daijavad, IBM.
- Hybrid Communications. Gaurav Bansal, Toyota InfoTechnology Center, USA, Thierry E., Klein, Nokia, and Sanjeev Athalye, Qualcomm Technologies, Inc.
- Sensing and Perception. Roger Berg, DENSO International America, Inc., and Michael Maile, Daimler AG.
- Security Innovation, Technologies for Data Ownership and Privacy. Angelos Amditis, Institute of Communication and Computer Systems, and Jonathan Petit, Security Innovation, Inc.

Summary of Key Takeaways and Lessons Learned from the Breakout Discussion

- Participants discussed opportunities for synergy related to maps, positioning, and the infrastructure through the use of crowdsourcing and road features. Another opportunity for synergy focused on hybrid communications of difficult use cases, such as collaborative driving in the long and short term.
- Speakers and participants discussed that the perspectives of deep learning may enable the use of low-cost sensors and provide the opportunity to leverage tremendous computational resources.
- Participants and speakers also discussed that data ownership and privacy are essential engineering and policy considerations.

Moving Forward

- Participants discussed that developing a summary of contemporary research directions and merging results would be beneficial.
- Participants also favored continuing the discussion at the 2017 AVS, including providing more time to explore and catalog synergisms between enabling technologies in more detail.

Breakout Session 6: Safety Assurance

Herman Winner, *Technische Universität Darmstadt, Germany*

Summary of the Meeting Agenda

This session included two panels and an opening discussion with participants.

The following presentations were part of the first panel:

- Dependability and Verification for Self-Driving Cars: The Drive Me Approach. Jonas Nillsson, Volvo Car Corporation, Gothenburg.
- Concerning Safety Assurance on Automated Vehicle: Results and Discussion Based on the Projects in Japan. Naohisa Hashimoto, National Institute of Advanced Industrial Science and Technology, Japan.
- Safety Assurance Based on an Objective Identification of Scenarios: One Approach of the PEGASUS Project. Walther Wachenfeld, Technische Universität Darmstadt, Germany.
- Developing and Assessing Automated Driving. Lutz Eckstein, RWTH Aachen University, Germany.

The following presentations were part of the second panel:

- Establishing Trust in Autonomous Vehicles: An Aerospace Perspective. Tim Allan Wheeler, Stanford University.
- Driving Autonomous Vehicles to Safety. Nidhi Kalra, RAND.
- Functional Validation and Performance Assessment of Automated Truck Platoons in Controlled Environments. Marcos Pillado, Applus IDIADA, Spain.

- Challenges in Autonomous Vehicle Testing and Validation. Micahael Wagner, Carnegie Mellon University.
- Applicability of Lessons Learned from Aviation Safety Management System for Automated Vehicles. Andrew Lacher, The MITRE Corporation.

Summary of Key Takeaways and Lessons Learned from the Breakout Discussion

- Some of the topics discussed during the first panel included that safety assurance implies a lot of redundancy and that safety-related tasks must be clearly divided between the driver and the autopilot mode. Speakers and participants discussed the strong interdependencies between levels of automation and safety assurance approaches. It was suggested by some participants that scenario-based testing methodology will need to be deployed step-by-step. Other topics discussed focused on the role of the driver at different levels of automation and how data for safety assurance can be collected commonly and shared for safety design.
- Some of the topics discussed during the second panel included that a scientific, unified framework to optimize and evaluate the safety of CAVs will lead to trust in AD. It was also suggested by some participants that it is not possible to wait for a perfectly safe system. Beginning with reasonable safety in order to save lives was suggested as a reasonable approach. Testing machine learning of automated drivers subsystems for robustness was discussed by participants. Some participants noted that lessons can be learned from aviation on their safety management processes.
- Other discussions focused on how learning systems could be tested and the benefits of monitoring unknown situations. Methods to communicate the fact that the AD may not be tested as perfect before introduction were also discussed by participants. In addition, aviation safety and road safety were compared concerning risk management and their financial resources for safety.

Moving Forward

- Participants suggested that providing open databases for test scenarios would be beneficial.
- Participants discussed that developing a safety management system similar to the model used in aviation would be of use.
- Participants suggested that research examining a scientifically accepted validation methodology was needed.
- Research on how to communicate safety expectations to the public was suggested by some participants.

Breakout Session 19: Cyber Security and Resilience Challenges and Opportunities for Self-Driving Vehicles

Edward Fok, *Federal Highway Administration*

Summary of the Meeting Agenda

This session focused on three major objectives. The first objective was identifying and discussing the security, safety, and resilience goals of self-driving vehicles. The second objective was identifying and discussing the challenges and opportunities the automotive kill chain presents for self-driving vehicles. The third objective was identifying and discussing the challenges and opportunities in an integrated automated transportation network.

The session included five presentations followed by a discussion of research opportunities:

- Attack Schenario on Automated Vehicle Platoon. Ryan Gerdes, Utah State University.
- Integration of Self Driving Vehicles into the Traffic Management and Internet of Things Ecosystems. Brian Russell, Leidos.
- Security for Connected Automated Driving. Tim Leinmuller, Denso.
- Assessing the Risk of Cyber Infrastructure Vulnerabilities: From Legacy to Next Generation. Kenneth Perrine, University of Texas at Austin.
- Prevent Accidents and Save Fuel by Connecting Trucks. Chuck Price, Peloton Technologies, Inc.

Summary of Key Takeaways and Lessons Learned from the Breakout Discussion

- Speakers and participants discussed the opportunities for vehicles and the roadway system to cooperate to improve misbehavior detection to improve the security of the overall network.
- Participants noted that the subject matter related to cyber security was extremely broad, with a number of topics for more detailed discussions identified.
- The assignment of threat actors or the potential attackers were discussed by the speakers and session participants. Threat actors may include kids hacking into the system to change traffic signals, activists protesting a project, criminals, and international terrorist organizations.
- Speakers and participants discussed attack scenarios, including the failure mode in highly complex systems. Platoon telematics attacks, which can cause a collision in a high-speed platoon, were discussed, as were random area attacks against a vehicle system.

Moving Forward

- Research on writing specifications for automated systems that are secure and useable to a wide audience was identified as a need by participants.
- Examining the appropriate balance between security and usability, including the difference between what the consumer expects and the need of the technical–engineering

community was another research topic discussed in the group. It was suggested that systems can be made very secure, but if it is too complex, people will not use it.

- Participants suggested that investigating alternatives to public key infrastructure would be beneficial.
- Examining methods to enable and measure crowdsourced data, either public or private, to improve misbehavior detection and to improve resiliency was another research topic suggested by participants.
- Developing and maintaining a sufficiently useable list of potential attacks, their severity, and their probability to enable an effective discussion about possible responses was identified by participants as a needed research topic.
- Participants suggested that research developing methods to gauge the effectiveness of specific security or resiliency solutions would be beneficial.
- Participants discussed the need for an open security and/or resilience research platform for AVs and if AUTOSAR could be updated to meet this need.

OPERATIONAL ISSUES FOR AUTOMATED VEHICLES

Breakout Session 13: Design and Operational Challenges and Opportunities for Deploying Automated Vehicles on Freeways and Managed Lanes

Chris Poe, *Texas A&M Transportation Institute*

Summary of the Meeting Agenda

This breakout session evolved from the 2015 session on CAVs with transportation management systems and transportation management centers. The TRB Freeway Operations Committee, Managed Lane Committee, and ITS Committee helped organize the session. Updates were provided on the Freeway Operations Committee activities including an e-circular on CAVs and Freeway Operations and a new section on emerging technologies in the *Freeway Operations and Management Handbook*. The following speakers provided updates on related projects and research:

- Ray Derr, TRB, provided an overview of *NCHRP 20-102: Impacts of Connected Vehicles and Automated Vehicles on State and Local Agencies*, including a project examining CAVs on dedicated lanes.
- Bob Ferlis, FHWA, provided an update on U.S. DOT projects underway at the Turner-Fairbanks Highway Research Center.
- Joe Rosa, California Department of Transportation (Caltrans), discussed lessons learned from managed lanes that are applicable to CAV dedicated lanes.
- Steve Kuciemba, WSP–Parsons Brinkerhoff, and Ali Zaqhari, Caltrans, described case studies from California, Florida, and Michigan, including autonomous vehicle operations on shoulder running lanes, the safety pilot model deployment–Ann Arbor connected vehicle test environment, the I-710 truck lanes, and the Florida driver-assisted truck platooning and Tampa CV pilot.

Summary of Key Takeaways and Lessons Learned from the Breakout Discussion

- One of the lessons learned from the breakout discussion was that the operating agencies have spent many years to reach agreement with stakeholders on the guidelines to operate managed lanes and to educate the public on their use. Participants discussed the need to involve those stakeholders in considering CAVs use of managed lanes.
- Participants also discussed the possible traffic impacts to high-occupancy vehicle (HOV) and managed lanes from CAV use, especially those facilities that are at or near capacity. Some participants voiced concerns that sufficient models and tools to evaluate these issues are not available. Some participants drew comparisons to concerns with current hybrid vehicle use of HOVs and managed lanes.
- Participants discussed requirements for CAV access, including the importance of evolving federal policy to allow CAVs on HOVs and managed lanes. Participants discussed operating in a mixed environment with CAVs and legacy vehicles, and possible pricing of CAVs, including if there is enough benefit to allow CAVs.
- The need for connectivity to TMCs, infrastructure for V2I applications, and IT support for communications, data, security, and back-end infrastructure was suggested by participants. The level of mapping needed for use of HOV and managed lanes, including access points, was also identified by participants as an important topic.
- The staffing resources needed to design, build, operate, and maintain the types of facilities were discussed by participants. The business models to staff these types of projects were considered by participants.

Moving Forward

- Participants suggested that international technology transfer would be beneficial on this topic.
- Developing a session for the 97th Annual Meeting of the TRB in 2017 or the 2017 AVS on modeling tools for CAVs on freeways and managed lanes were suggested as a needed follow-up activity. Topics to be included focused on the status of current tools, current research and development projects, and future research needs.
- Conducting deployment scenarios and table-top exercises for new or converted managed lane for CAVs was a related activity item. The TRB Managed Lanes Committee has a workshop scheduled for the 97th Annual Meeting of the TRB in 2017, which could include exploring deployment scenarios. It could also be incorporated into a session at the 2017 AVS working with multiple TRB committees.
- Future research identified by participants included sharing policy ideas with the policy session, examining what a transition to dedicated lanes for CAVs would include, and exploring elements to include in the next generation of transportation management system that utilizes CAVs.

Breakout Session 20: Infrastructure, Work Zones, and Digital Infrastructure

Paul Carlson, *Texas A&M Transportation Institute*
Carl Anderson, *U.S. Department of Transportation*

Summary of the Meeting Agenda

The major topics discussed in this session focused on AVs and the physical infrastructure, work zones, and digital infrastructure. 2016 was the third year a session on infrastructure has been included in the AVS. This year work zones and the digital infrastructure were added to the session.

- The physical infrastructure panel was moderated by Paul Carlson, TTI. Jina Wang, Carnegie Mellon University, discussed smartphone, sensors, and machine learning applied to inventory and condition assessment of physical infrastructure elements, primarily for use in asset management programs. Hideki Hada, Toyota, described machine vision interaction with traffic control devices. Paul Carlson, TTI, and Chris Davies, Potters Industry, described two ongoing research projects focused on identifying the road marking performance needs for camera-based vision systems. Scott Kuznicki, TOXCEL, presented examples of ambiguous infrastructure, when signals and pavement markings do not make sense to drivers or machines.
- The work zone panel moderated by Jerry Ullman, TTI, included four speakers. Colin Castle, Michigan DOT, and Jiaming Ma, Texas DOT provided updates on the connected work zone demonstration projects underway in Michigan and Texas. Ross Sheckler, iCone Products, presented examples of work zone information currently available on various mapping applications and the needs for effective CV applications. Jerry Ullman, TTI, discussed the need to obtain work zone locations and effects on capacity in real time, with a realization that it is extremely challenging.
- The digital infrastructure panel was led by Carl Anderson, U.S. DOT. Ryota Shirato, Nissan, provided an update on the second year of Japan SIP-adus, which focuses on a cross-ministerial program to develop methods of acquiring and providing dynamic data. Maxime Flament, ERTICO, described the regulatory framework developed in the EU to guide member states on how dynamic data should be provided to road users. Ahmed Nasr, HERE, discussed SENSORIS, which is a draft standard that allows a common method for moving vehicle data to the cloud. Russel Shield, Ygomi LLC, discussed probe data for AD. He suggested that high-definition maps meet the needs of drivers who are still in control with Level 1 and Level 2 applications, but that an entirely different approach is needed for AD.

Moving Forward

- Noting the progress that has been made over the past 3 years, participants suggested that continuing the dialogue on physical infrastructure, work zones, and digital infrastructure needs was important.
- Participants identified a number of follow-up activities associated with TRB committees. One suggestion was to host a session at the 97th Annual Meeting summarizing the key elements of the breakout session. Topics identified for further discussion at the TRB Traffic Control Device Committee included developing a call for papers for the 98th Annual Meeting in

2018 and developing research problem statements focusing on a collaborative review of infrastructure standards and a framework for defining road readiness for automated vehicles. Expanding the TRB Symposium on Visibility and Traffic Control Devices, which occurs every 2 years, to include some of the infrastructure topics discussed in the breakout session was another issue mentioned by participants.

Breakout Session 21: Traffic Flow of Connected Automated Vehicles

Xiaopeng (Shaw) Li, *University of South Florida*

Summary of the Meeting Agenda

This session included two key note presentations, three shorter presentations, and a general discussion with all participants:

- The two keynote presentations were given by Hami Mahmassani, Northwestern University, and Pravin Varaiya, University of California, Berkeley. Mahmassani discussed AVs and traffic physics. Varaiya described the potential improvement in the capacity of intersections from platooning vehicles.
- The shorter presentations featured three speakers addressing current projects and activities. Osman Altan, FHWA, discussed recent research projects and experiments at the Turner-Fairbank Highway Research Center, including lane management, speed navigation, AD, and truck platooning. Simeon Calvert, TNO, described a recent review of research gaps in traffic flow theory and modeling for consideration of the potential impacts of AD in mixed traffic. Jan-Niklas Meier, CAMP, highlighted recent projects investigating and testing of CACC and CACC technology. The five speakers participated in a panel discussion, which included questions and comments from the audience.

Summary of Key Takeaways and Lessons Learned from the Breakout Discussion

- Participants discussed the ability to use ACC to reduce reaction time at intersections.
- Methodologies to help advance traffic flow theory and the deployment of CAVs were discussed by participants. Using simulation as a virtual testbed was suggested by some participants.
- Using traffic worthiness as a criterion to design and verify CAV algorithms was also suggested by participants. Possible performance measures identified by participants included measures of safety throughout (capacity), stability (local and global), flow breakdown (reliability), and sustainability.
- Implications related to traffic flow modeling were discussed by participants in the session. Participants suggested the need for flexibility in incorporating different communication latencies. Fundamental issues associated with human driver behavior and vehicle interactions were noted as important by participants.

Moving Forward

- The need to conduct field experiments and obtaining ground truth data to validate model assumptions was one action item identified by participants.
- Participants discussed that inexpensive and accurate positioning technology is critical to CACC implementation, but that it is currently not available. More research on this topic was suggested as needed by participants, including examining the ability to fully utilize DSRC channels.
- Research topics identified by participants included examining the potential for CACC platoons to be broken by nonconnected vehicles, and the impact when this situation occurs. Exploring the operating scenarios of various combinations of CAVs, AVs, and legacy vehicles was another area of possible research.
- Assessing the impact of infrastructure and weather on traffic flow models and examining the impact of electric vehicles were two other areas of possible research identified by participants.

Breakout Session 8: Traffic Signal Control with Connected and Automated Vehicles

Larry Head, *University of Arizona*

Summary of the Meeting Agenda

This session focused on the impact on traffic signal control systems from a very low percentage of CAVs to a higher percentage of CAVs and eventually to fully autonomous transportation systems when traffic signal controls may no longer be needed.

The breakout session included the following four presentations:

- Multimodal Considerations in CAV Traffic Control. Larry Head, University of Arizona.
- Next Generation Traffic Signal Systems with CAVs. Henry Liu, University of Michigan.
- Core Technologies for Cooperative Vehicle Adaptive Signal Control Systems (CoVASS). Young-Jun Moon, Korea Transport Institute.
- Intersection Vehicle Infrastructure Cooperative Traffic Management and CAVs. Wei-Bin Zhang, California PATH Program.

Summary of Key Takeaways and Lessons Learned from the Breakout Discussion

- The speakers discussed many of the same fundamental questions related to the need for traffic signals in the future with a fully autonomous system and the transition during market adoption of CAVs. Speakers noted that traffic signals were initially deployed for efficiency purposes, rather than safety purposes. Accommodating large volumes of CAVs at intersections may still require traffic signals. The speakers suggested that traffic signals will continue to be needed during the transition phase and well into the future.

Moving Forward

The breakout group participants identified the following general categories of research related to traffic signals and the deployment of CAVs:

- Participants discussed the need for research examining network-level control considerations, which are more than a collection of intersections. For example, some participants suggested that data available from CAVs can be used to analyze heterogeneous path flows.
- Participants suggested that research on intersection user capabilities and characteristics would be beneficial. Intersection users include vehicles, pedestrians, trucks, buses, bicyclists, and motorcyclists.
- Exploring institutional and social issues associated with changes to higher signals to accommodate a much more diverse fleet mix with CAVs and non-CAVs was noted as important by participants
- Participants discussed the need to assess traffic flow theory related to traffic signals. Changes in vehicle behaviors, including saturation flow, headway, acceleration, start-up lost time, and vehicles in the yellow phase were identified for consideration.
- Participants identified application scenarios that could be examined in research, including managed lanes for CAVs, multimodal applications, integration of apps, speed harmonization, and eco-driving.
- Developing new traffic signal control algorithms and strategies based on the improved data available from CAVs was another research topic suggested by participants. Elements to consider included trajectory control, multimodal concerns, priority, path-based, and vehicle dynamics.
- The need for research examining human factors issues at intersections was discussed by participants, including passenger and driver limits to acceleration and gaps.
- The potential need for infrastructure adaptation was discussed by participants. Research examining potential geometric changes and needed infrastructure modifications was suggested as beneficial by some participants.
- Exploring alternative paths from today to the next generation CAVs and traffic signals was discussed by participants. One suggested research topic was developing levels of automation for traffic signals, similar to the vehicle automation levels.
- Participants discussed the impact of shared mobility on traffic signal control. It was suggested that large fleets of vehicles could agree to work cooperatively with the infrastructure to improve overall operations. These vehicle fleets may include Uber, Lyft, buses, and other services.

Poster Sessions

RESEARCH POSTER SESSION

The Effects of a Change in Driving Environment on Drivers' Ability to Anticipate Latent Hazards during a Take-Over Request Following Level 3 Automation

- Ravi Agrawal, University of Massachusetts Amherst;
- Timothy J. Wright, University of Massachusetts Amherst;
- Siby Samuel, University of Massachusetts Amherst;
- Avinoam Borowsky, Ben-Gurion University of the Negev;
- Shlomo Zilberstein, University of Massachusetts Amherst; and
- Donald L. Fisher, University of Massachusetts Amherst.

Estimate of Fuel Consumption and GHG Emission Impact from an Automated Mobility District

- Yuche Chen, NREL;
- Stanley Young, NREL;
- Xuewei Qi, University of California, Riverside; and
- Jeffrey Gonder, NREL.

Impact of In-Vehicle Display Position on Motion Sickness in Automated Vehicles: An On-Road Study

- Cyriel Diels, Coventry University;
- Jelte E. Bos, TNO Perceptual and Cognitive Systems;
- Vrije Universiteit Amsterdam, Netherlands;
- Katharina Hottelart, Valeo; and
- Patrice Reilhac, Valeo.

Putting the Brakes on Autonomous Vehicle Control: Considerations for Implementation

- Kelly Funkhouser, University of Utah; and
- Frank Drews, University of Utah.

Critical Importance of Standards for the Communication Between Autonomous Vehicles and Humans

- Colleen Emmenegger, Design Lab, University of California San Diego;
- Malte Risto, Design Lab, University of California San Diego;
- Ben Bergen, Cognitive Science Department, University of California San Diego;
- Donald L. Norman, Design Lab, University of California San Diego; and
- Jim Hollan, Design Lab, University of California San Diego.

Fusion of Information from Local Sensors and V2X-Communicated Data for Automated Driving

- Yaser Fallah, West Virginia University;
- Ahmed Cheikh Sidiya, West Virginia University;
- Masoumeh Kalantari, West Virginia University;
- Gaurav Bansal, Toyota Info Technology Center;
- Xin Li, West Virginia University; and
- Takayuki Shimizu, Toyota Info Technology Center.

SMOOTH: Improved Short-Distance Mobility for a Smarter City

- Xiangjun Fu, Department of Electrical and Computer, The Ohio State University;
- Michael Vernier, Department of Electrical and Computer Engineering, The Ohio State University;
- Arda Kurt, Department of Electrical and Computer Engineering, The Ohio State University;
- Keith Redmill, Department of Electrical and Computer Engineering, The Ohio State University; and
- Umit Ozguner, Department of Electrical and Computer Engineering, The Ohio State University.

New Cockpit Concept for SAE Level 3 Automation: Evaluation in a Car Simulator

- Katharina Hottelart, Valeo;
- Patrice Reilhac, Valeo; and
- Christopher Nowakowski, Valeo.

Capacity Impacts and Optimal Geometry of Automated Cars' Parking Facilities

- You Kong, Southwest Jiaotong University;
- Scott LeVine, Southwest Jiaotong University;
- Xiaobo Liu, Southwest Jiaotong University; and
- Lianyu Chu, Southwest Jiaotong University.

Anticipating the Emissions Impacts of Autonomous Vehicles Using the MOVES Model

- Jun Liu, Center for Transportation Research, The University of Texas at Austin;
- Kara M. Kockelman, Department of Civil, Architectural and Environmental Engineering, The University of Texas at Austin; and
- Aqshems Nichols, Department of Civil, Architectural and Environmental Engineering, The University of Texas at Austin.

Tracking a System of Shared Autonomous Vehicles Across the Austin, Texas, Network Using Agent-Based Simulation

- Jun Liu, Center for Transportation Research, The University of Texas at Austin;
- Kara M. Kockelman, Department of Civil, Architectural and Environmental Engineering, The University of Texas at Austin;
- Patrick M. Boesch, IVT, ETH Zurich; and
- Francesco Ciari, IVT, ETH Zurich.

Integrated Control for Platooning in Truck Partial Automation

- Xiao-Yun Lu, California PATH Program, University of California, Berkeley;
- Steven E. Shladover, California PATH Program, University of California, Berkeley;
- Stefan Bergquist, Volvo Group Technology;
- Aravind Kailas, Volvo Group Technology;
- Deborah Thompson, Volvo Group Technology;
- Matt Hanson, California Department of Transportation; and
- Osman Altan, Federal Highway Administration.

Scenario-Based Approach for Anticipating the Impacts of AVs on Transportation and Urban Form

- AnnaLisa Meyboom, Director, TIPS Lab, Associate Professor, School of Architecture and Landscape Architecture, University of British Columbia; and
- Sara Costa Maia, University of British Columbia.

Naturalistic Study of Truck-Following Behavior

- Emily Nodine, Volpe;
- Andy Lam, Volpe;
- Mikio Yanagisawa, Volpe;
- Dr. Wassim Najm, Volpe; and
- Govindarajan Vadakpat, Federal Highway Administration.

Estimating Energy and Mobility Benefits from an Eco-Approach and Departure System for Electric Vehicles

- Xuewei Qi, Electrical and Computer Engineering Department and Center for Environmental Research and Technology, University of California at Riverside;
- Guoyuan Wu, Center for Environmental Research and Technology, University of California at Riverside;
- Kanok Boriboonsomsin, Center for Environmental Research and Technology, University of California at Riverside; and
- Matthew Barth, Electrical and Computer Engineering Department and Center for Environmental Research and Technology, University of California at Riverside.

Vehicle Automation and the Driver Age: Human Factors Implications

- Brandon Smith;
- Charles Dyer, Department of Computer Science, University of Wisconsin–Madison;
- Madhav Chitturi, TOPS Laboratory, Department of Civil and Environmental Engineering, University of Wisconsin–Madison; and
- John D. Lee, Department of Industrial Engineering, University of Wisconsin–Madison.

Centimeter-Level Highway Speed Localization at Night in a Snowstorm Using Localizing Ground-Penetrating Radar

- Byron Stanley, MIT Lincoln Laboratory;
- Jeffrey Koechling, MIT Lincoln Laboratory; and
- Matthew Cornick, MIT Lincoln Laboratory.

Modeling the Impact of Automated Driving: Private Automated Vehicle Scenarios for Germany and the United States

- Stefan Trommer, German Aerospace Center;
- Lars Kröger, German Aerospace Center; and
- Tobias Kuhnimhof, German Aerospace Center.

Developing an Eco-Cooperative Adaptive Cruise Control System

- Guoyuan Wu, Center for Environmental Research and Technology, University of California at Riverside;
- Peng Hao, Center for Environmental Research and Technology, University of California at Riverside;
- David Kari, Electrical and Computer Engineering Department, University of California at Riverside;
- Kanok Boriboonsomsin, Center for Environmental Research and Technology, University of California at Riverside; and
- Matthew Barth, Electrical and Computer Engineering Department and Center for Environmental Research and Technology, University of California at Riverside.

Eco-Approach and Departure Strategy in Traffic Under Mixed Connected Vehicle Environment

- Fei Ye, Electrical and Computer Engineering Department, University of California at Riverside;
- Peng Hao, Center for Environmental Research and Technology, University of California at Riverside;
- Guoyuan Wu, Center for Environmental Research and Technology, University of California at Riverside;

- Kanok Boriboonsomsin, Center for Environmental Research and Technology, University of California at Riverside; and
- Matthew Barth, Electrical and Computer Engineering Department and Center for Environmental Research and Technology, University of California at Riverside.

SPECIAL INTEREST POSTER SESSION

Legislation in Sweden Regarding Self-Driving Vehicles

- Jonas Bjelfvenstam (inquiry chair);
- Kristina Andersson (inquiry secretary);
- Ann-Cathrine Wikstrom (inquiry secretary);
- Jonas Bjelfvenstam, Regeringskansliet (inquiry chair);
- Kristina Andersson, Regeringskansliet (inquiry secretary); and
- Ann-Cathrine Wikstrom, Regeringskansliet (inquiry secretary).

Mixed-Function Automation Naturalistic Driving Study

- Myra Blanco, Virginia Tech Transportation Institute;
- Sheldon Russel, Virginia Tech Transportation Institute;
- Vikki Fitchett, Virginia Tech Transportation Institute;
- Tammy Trimble, Virginia Tech Transportation Institute; and
- Paul Rau, National Highway Traffic Safety Administration.

NCHRP Project 20-102: Impacts of Connected Vehicles and Automated Vehicles on State and Local Transportation Agencies

- Ray Derr, NCHRP, The National Academies of Sciences, Engineering, and Medicine.

Infrastructure for Automated Driving: Mind Map for Future Research Directions

- Haneen Farah, Transport and Planning, Faculty of Civil Engineering and Geosciences, Delft University of Technology, Netherlands;
- Bart van Arem, Transport and Planning, Faculty of Civil Engineering and Geosciences, Delft University of Technology, Netherlands; and
- Sandra M.J.G. Erkens, Pavement Engineering, Faculty of Civil Engineering and Geosciences, Delft University of Technology and Ministry of Infrastructure and the Environment, Rijkswaterstaat Water, Traffic and Environment, Netherlands.

Simulation-Based Traffic Management System for Connected and Autonomous Vehicles

- Pawel Gora, Faculty of Mathematics, Informatics and Mechanics, University of Warsaw.

Will Cars Drive Themselves More or Less Than Humans Do?
Estimating Travel Behavior Changes with Autonomous Vehicles

- Kevin Johnson, Fehr & Peers;
- Ron Milam, Fehr & Peers; and
- Jeff Pierson, Fehr & Peers.

Heavy Truck Cooperative Adaptive Cruise Control Testing and Demonstration Plans

- Aravind Kailas, Volvo Group;
- Stefan Bergquist, Volvo Group;
- Deborah Thompson, Volvo Group;
- Steven E. Shladover, California PATH Program, University of California, Berkeley;
- Xiao-Yun Lu, California PATH Program, University of California, Berkeley;
- Matt Hanson, Caltrans; and
- Osman Altan, FHWA.

Review of Federal Motor Vehicle Safety Standards (FMVSS) for
Automated Vehicles: Identifying Potential Barriers and Challenges
for the Certification of Automated Vehicles Using Existing FMVSS

- Anita Kim, U.S. DOT;
- David Perlman, U.S. DOT;
- Dan Bogard, U.S. DOT; and
- Ryan Harrington, U.S. DOT.

Impacts of Road Grade on Cooperative Adaptive Cruise Control Vehicle Platoons

- Danjue Chen, Department of Civil and Environmental Engineering, University of Wisconsin–Madison;
- Soyoung Ahn, Department of Civil and Environmental Engineering, University of Wisconsin–Madison;
- Madhav Chitturim, Department of Civil and Environmental Engineering, University of Wisconsin–Madison; and
- David Noyce, Department of Civil and Environmental Engineering, University of Wisconsin–Madison.

Automated and Connected Vehicle Research at The Ohio State University

- Arda Kurt, Department of Electrical and Computer Engineering, The Ohio State University;
- Keith Redmill, Department of Electrical and Computer Engineering, The Ohio State University; and
- Umit Ozguner, Department of Electrical and Computer Engineering, The Ohio State University.

Class 8 Tractor Trailer Platooning Effects, Impacts, and Improvements

- Michael Lammert, National Renewable Energy Laboratory;
- Jeff Gonder, National Renewable Energy Laboratory;
- Ken Kelly, National Renewable Energy Laboratory;
- Kamiz Salari, Lawrence Livermore National Laboratory; and
- Jason Ortega, Lawrence Livermore National Laboratory.

UACT—Universal Automated Community Transport

- Elizabeth Machek, Volpe;
- Joshua Cregger, Volpe;
- Sean Peirce, Volpe;
- Joseph Stanford, Volpe;
- Dan Bogard, Volpe;
- Alexandra Oster, Volpe;
- Jonathan Frazier, Volpe;
- Yousuf, Mohammed, FHWA; and
- Kevin Dopart, ITS JPO, U.S. DOT.

Safeguarding Privacy While Sharing: The Prospects of an Autonomous Fleet

- Lauren McCarthy, George Mason University; and
- David Morar, George Mason University.

A Drive into Understanding Consumer Mental Models and Trust Regarding Driver Assistance Technologies

- Ashley McDonald, University of Iowa, Transportation and Vehicle Safety Policy Research Program, Public Policy Center;
- Daniel McGehee, University of Iowa, Transportation and Vehicle Safety Policy Research Program, Public Policy Center;
- Michelle Reyes, University of Iowa, Transportation and Vehicle Safety Policy Research Program, Public Policy Center; and
- Cheryl Roe, University of Iowa, Transportation and Vehicle Safety Policy Research Program, Public Policy Center.

Road Transport Automation in Finland: Roadmap, Action Plan, and Ongoing Initiatives

- Eetu Pilli-Sihvola, Finnish Transport Safety Agency; and
- Risto Kulmala, Finnish Transport Safety Agency.

Vehicle Automation and the Driver Age: Human Factors Implications

- Anuj Pradhan, University of Michigan;
- Lisa Molnar, University of Michigan; and
- Lindsey Ryan, University of Michigan.

Applied Robotics for Installation and Base Operations

- Joseph Putney, Robotic Research, LLC;
- Edward Mottern, Robotic Research, LLC; and
- Janet Hughes, Robotic Research, LLC.

A Programme of Connected and Automated Vehicles Research in the U.K. Smart Mobility Living Lab

- Nick Reed, TRL.

Analysis Methods for Assessing Energy Impacts of Connected and Automated Vehicles at the U.S. National Level

- Thomas Stephens, Argonne National Laboratory;
- Jeffrey Gonder, National Renewable Energy Laboratory;
- Changzheng Liu, Oak Ridge National Laboratory;
- Joshua Auld, Argonne National Laboratory;
- Yuche Chen, National Renewable Energy Laboratory;
- Zhenhong Lin, Oak Ridge National Laboratory;
- Kouros Mohammedian, University of Illinois;
- James Li, Oak Ridge National Laboratory; and
- Ramin Shabanpour, University of Illinois at Chicago.

Future Event Recorders for Automated Vehicles

- Ryan Yee, Exponent; and
- Carmine Senatore, Exponent.

APPENDIX A: BREAKOUT SESSION 1

Public Transport and Shared Mobility

Organizers

- Gary Hsueh, Arup;
- Daniel Fagnant, General Motors;
- James Fishelson, University of Michigan;
- Lauren Isaac, WSP–Parsons Brinkerhoff;
- Jessica Lazarus, University of California, Berkeley Transportation Sustainability Research Center;
- Matthew Lesh, Local Motors;
- Rachel Liu, New Jersey Institute of Technology;
- Sam Lott, Texas Southern University;
- Shannon Sanders McDonald, Southern Illinois University Carbondale;
- Nadereh Moini, New Jersey Sports and Exposition Authority;
- Susan Shaheen, University of California, Berkeley Transportation Sustainability Research Center;
- Adam Stocker, University of California, Berkeley Transportation Sustainability Research Center;
- Tom Voege, OECD; and
- Stanley Young, NREL.

Reporters

- Jessica Lazarus, University of California, Berkeley; and
- Katherine Turnbull, TTI.

SESSION FOCUS

This breakout group sponsored two sessions examining how vehicle automation technology could be harnessed across public transport and shared mobility services to provide mobility for all segments of the population. Speakers and participants discussed the potential for vehicle automation to disrupt traditional transit systems, new and different types of market-driven and publicly run frameworks that may emerge, and the performance of future public transport. PowerPoint presentations of most of the presentations are available on the AUVSI website at http://www.automatedvehiclessymposium.org/program/breakouts/home.

SESSION SUMMARY

Quick Bursts A: Updates on Research, Projects, Pilot Programs, and Testing Sites

Susan Shaheen, *University of California, Berkeley, moderator*

Speakers	Summary of Presentation and Question and Answer Session
Susan Shaheen, University of California, Berkeley, Transportation Sustainability Research Center	Susan Shaheen discussed a recent study examining the use of car2go in five North American cities, which included surveys of members. The study presents disaggregated level data on vehicle use in a one-way vehicle use model. The study found that 2% to 5% of car2go members in the five cities sold a vehicle due to car2go and 7% to 10% did not acquire a vehicle due to car2go. The average VMT reduction per household after joining car2go was 11%.
Randell Iwasaki, Contra Costa Transportation Authority (CTTA)	Randell Iwasaki described GoMentum Station, which is a 5,000-acre facility for CV and AV innovation and commercialization. The overarching goals of GoMentum are to foster economic growth and job creation, efficient mobility, a healthier environment, and enhanced safety. EasyMile is testing automated pods at GoMentum. The pods will be deployed in a pilot at the Bishop Ranch business park in San Ramon to provide first-mile/last-mile service to Bay Area Rapid Transit (BART) stations.
Justin Holmes, Zipcar	Justin Holmes described the development of Zipcar, which was founded in Boston in 2000. The Zipcar mission is to enable simple and responsible urban living. The broader social vision is a world where car sharers outnumber car owners. There are currently nearly 1 million fee-paying members worldwide. The typical system is 800 to 900 vehicles in urban areas. Studies by the University of California, Berkeley, indicate a reduction of approximately 1 ton of CO_2 per member per year and a 27% to 43% reduction in VMT per household. He suggested that consumer attitudes are changing in urban areas with more interest in shared mobility and enabling technologies.
Robbert Lohmann, 2getthere	Robbert Lohmann discussed automated transit systems, including automated people mover systems, automated transit networks, and shared autonomous vehicles. He suggested safety and person throughput were two important characteristics of AV systems. He noted that a study is being conducted in Singapore exploring the first application.

Quick Bursts B: Updates on Research, Projects, Pilot Programs, and Testing Sites

Susan Shaheen, *University of California, Berkeley, moderator*

Speakers	Summary of Presentation and Question and Answer Session
Adriano Alessandrini, University of Florence	Adriano Alessandrini expanded on his presentation in the general session on the EU CityMobil2 pilots. He stressed the importance of addressing legal and policy issues related to driverless shuttles and AVs. He also noted the need to select the right vehicles for the appropriate performance.
Chris Kopp, HNTB Corporation	Chris Kopp described a study conducted in partnership with the Florida DOT, which applied an economic model to evaluate the impacts of shared automated taxis from society, transit agency, taxpayer, and customer perspectives. The results indicated that fixed-route service is more advantageous to society, even when ridership is low; that shared automated taxis provide a method for agencies to increase overall system effectiveness; and as fares for the shared automated taxis decreased they become more attractive from a taxpayer perspective and to customers. Continuing to focus fixed-route transit in the densest corridors was another study finding.
Chris Augenstein, Santa Clara County Valley Transportation Authority (VTA)	Chris Augenstein described the agencies involved in research and pilots including an on-demand service pilot implemented in early 2016. VTA is also focusing on redesigning bus services to better serve the new BART line. A multimodal trip planner and virtual transit trip to acquaint new riders are currently being beta-tested.
Gary Hsueh, Arup	Gary Hsueh provided an update on the U.K. Autodrive Programme. He described the three pilot projects. The M1 Car Development Project focused on full-size automated vehicles operating in increasingly difficult test events. The Cities Programme included a national longitudinal public attitude survey, congestion simulations, and last-mile pilots in Milton–Keynes and Coventry. The L-SATS Development Project focused on designing pods for low-speed personal transportation in pedestrian areas. He noted that collaboration with the numerous consortium members takes time but is key to success.

Presentations: Program Updates and Funding Opportunities

James Fishelson, *moderator*

Speakers	Summary of Presentation and Question and Answer Session
Vincent Valdes, FTA	Vincent Valdes described the FTAs MOD Sandbox project. He noted that key principles of MOD are traveler-centric and consumer-driven; data-connected and platform independent; and mode agnostic and multimodal system. The MOD Sandbox is intended to provide transit agencies with a safe space and funding to innovate and learn from pilots. Currently, 79 project proposals from 33 states with a total funding request of $59 million are being evaluated.
Kevin Dopart, U.S. DOT ITS JPO	Kevin Dopart described the Smart City Challenge. A total of 78 applications were received. Approximately half the proposals included some type of electric automated pod, vehicle, or shuttle. Seven cities were selected as finalists, with Columbus, Ohio, selected as the final winner. The Columbus proposal, which focused on underserved areas, includes an electric shuttle service providing first- and last-mile service to a commercial district and truck platooning on arterials. He noted the interest generated in communities throughout the country by the Smart City Challenges.
Michael McGurrin, Noblis	Michael McGurrin described the Accessible Transportation Technology Initiative, which is a joint U.S. DOT effort, co-led by FHWA and FTA, with support from the ITS JPO. He outlined activities in the first phase focused on exploratory and user needs research. Webinars, listening groups, and other methods were used to develop the User Needs Assessment: Stakeholder Engagement Report. The second phase will include innovation, prototype development, and testing of technologies to address identified needs.
Doug Gettman, Kimley-Horn and Associates, Inc.	Doug Gettman described one of the NCHRP 20-102 series, Impacts of Connected Vehicles and Automated Vehicles on State and Transportation Agencies projects, on the Impacts of Regulations and Policies on CV and AV Technology Introduction in Transit Operations. The project is examining the current public transit regulatory and policy landscape that could influence the introduction of CV and AV technologies in transit, identifying needed changes in regulations and policies to enhance the use of CV and AV technologies in transit, and identify the implications of CV and AV technologies for transit stakeholders.
Katherine Kortum, TRB	Katherine Kortum discussed activities underway at TRB related to CV and AV research. Examples she noted included the recently published *NCHRP Legal Research Digest 69: A Look at the Legal Environment of Driverless Vehicles*, the numerous projects underway in the NCHRP 20-102 series, and the Partners in Research Symposium in Detroit on October 31–November 1, 2016. She also highlighted related TCRP and Airport Cooperative Research Program projects, as well as numerous TRB committee activities.

Shark Tank: Thinking About Setting up a Pilot Program? What You Need to Succeed

Tom Voege, *ITF, OECD, moderator*

Speakers	Summary of Presentation and Question and Answer Session
Tom Voege, OECD	Voege discussed some of the lessons learned related to regulator frameworks, enabling infrastructure, humane elements, and establishing business cases from the initial EC-funded projects. He moderated the Shark Tank with participants Jonathan Matus, Zendrive, and Adriano Alessandrini, University of Florence. The judges included Randy Iwasaki, CCTA, Chris Augenstein, VTA, and Tom Voege, OECD

Integration of Public and Private Models

Dan Fagnant, *moderator*

Speakers	Summary of Presentation and Question and Answer Session
Barbara Laurenson, Metropolitan Transportation Commission (MTC)	Laurenson discussed carpooling and ridesharing apps. She noted that the San Francisco Bay Area has had a regional ride-matching system for decades. There are currently at least seven carpool matching and ridesharing apps in the region, which covers nine counties. The MTC is focusing on promoting and encouraging carpooling, regardless of the app, company, or approach being used.
Michael Scrudato, Munich Re	Scrudato discussed factors associated with technologies and shared-mobility services that may influence the automobile insurance industry. He outlined public–private partnership pilot projects, including the Washington State Transit Insurance Pool. As part of the pilot, four buses were equipped with CWSs, which are being evaluated. The system may be expanded to additional buses in the future.
Emily Castor, Lyft	Castor described Lyft and public transit. She noted that Lyft can help serve the first-and last-mile needs of public transit systems when geographies, population density, and services make sense. She highlighted the experience with the Livermore–Amador Valley Transit Authority. She suggested that automated shared-ride systems will make the service more affordable.
Matthew George, Bridj	George discussed the approach being used with Bridj, which he likened to creating pop-up bus routes to serve new and emerging markets and enhance existing services. He described the project in Kansas City. Individuals use a smartphone app to request a trip, which is served through vehicles making dynamic pick-ups and drop-offs within specific areas.
Susan Shaheen, University of California, Berkeley, Transportation Sustainability Research Center	Shaheen expanded on her earlier comments and highlighted studies examining the impacts of car sharing. She described a 2008 study of roundtrip car shares and a 2015 study examining one-way free-flowing car sharing. She also described a 2015 study of bike sharing in the United States, Canada, and Mexico.

What Is Public Transport in the Future?

Stan Young, *NREL/ATRA, moderator*

Speakers	Summary of Presentation and Question and Answer Session
John Mirisch, Mayor of Beverly Hills	Mirisch described opportunities for automated shuttle systems, especially servicing first-and last-mile connections from LRT, commuter rail, and busways. He suggested that transit-oriented development may become autonomous-oriented developments. He noted that automated shuttle systems would provide improved mobility for older and younger individuals, people with special needs, and other groups. He also noted the interest on the topic at the U.S. Conference of Mayors.
Stan Young, NREL/Advanced Transit Association	Young discussed the importance of public mobility in light of continuing global urbanization, as well as the energy efficiency gains of an electric, centrally managed fleet of vehicles. He described the sweet spot of automated, electric, and shared vehicles in providing public mobility. He highlighted the EU CityMobile2 projects, the pilot project under development at Bishop Ranch using two EasyMile shuttles, and other international pilots.
Sam Lott, Texas Southern University	Lott described the parallel paths of CVs and AVs and suggested they come together in a fully automated transit system. He discussed the use of hazard analysis, which focuses on identifying key hazards related to potential accidents, safety critical parameters, and approaches to address key issues. He reviewed the FTA Analysis Guidelines, the American Society of Civil Engineers 21 Hazard Resolution Process, and other related programs. He noted this approach could be used to identify CV and AV technology to resolve key transit safety issues.
Jerome Lutin, New Jersey Transit (retired)	Lutin described the captive rider and choice rider markets served by public transit. He discussed the potential impacts of AVs on transit and use of the technologies available at different levels of automation in transit. He noted the potential benefits to transit from lane keeping, precision docking, collision avoidance, and other technologies.
Mark Mindorff, DART	Mindorff discussed the potential impacts of CVs and AVs on paratransit services. He noted that technology, including AVs, have the potential to reduce operating costs for paratransit services. Other suggested benefits were using data analytics to predict trip cancelations, improving safety with crash notification systems, and smartphone apps to alert customers of approaching vehicles.

Workshop: Policy Implications and Research Needs for Public Transport and Shared Mobility

Will Baumgardner, *moderator*

Participants discussed a wide range of issues and research needs during the workshop. Topics covered focused on equity, increasing mobility across geographic and economic ecosystems, safety, and system efficiency.

SUMMARY

- Participants discussed that shared AVs will begin to blur the lines between public and private transportation. This approach could deliver efficient and affordable public transportation to meet societal needs—improving access to jobs and health care. Participants noted that it would not replace mainline public transportation, however.
- Participants discussed deployment opportunities for first-and last-mile connections, as well as addressing the needs of underserved populations and areas lacking quality transit service. Participants noted that these opportunities serve a much broader market.
- Participants suggested that cities and sites are different, so that shared AV deployments need to be tailored to varying technical, cultural, and legal contexts. The strong interest at the local level in testing and deploying shared AVs was noted by many participants.
- Participants discussed that pilot programs, enabled by public–private partnerships, are encouraging private shared services to adapt and expand their functionality to meet the needs of public transit users.
- It was noted by participants that there are many competitive federal funding opportunities that highlight the role of shared AVs in public transportation.
- Participants suggested that significant thought is required in setting and adapting policies at all levels of government to support the use of automation and shared AVs in public transportation.

MOVING FORWARD

Developing and adopting a traveler-centric approach to transportation planning that stresses equity, occupancy, and sustainability was suggested by participants as possible future research.

- Participants discussed the need to promote public–private partnerships that leverage the strengths unique to each sector.
- Participants noted that it would be beneficial to add flexibility in transport procurement processes to consider mobility as a whole.
- Participants suggested the developing of safety standards to enable the implementation of automated technology in public fleets was needed.
- Breakout group participants encouraged the use of pilot programs as a safe space for experimentation.
- Participants discussed the importance of measuring, documenting, and sharing best practices and impacts of automated shared mobility and public policy adaptations.

APPENDIX B: BREAKOUT SESSION 2

Law and Policy as Infrastructure

Organizers

- Ellen Partridge, United States DOT;
- Karlyn Stanley, RAND;
- Bryant Walker Smith, University of South Carolina School of Law;
- Susan Spencer, Susan Spencer & Associates;
- Austin Hensel, Florida DOT;
- Paul Hemmersbaugh, NHTSA;
- Jim Thiel, consultant;
- Catherine Curtis, FTA;
- Jesse Chang, NHTSA; and
- James Anderson, RAND.

Reporters

- Austin Hensel, Florida DOT; and
- Karlyn Stanley, RAND.

SESSION FOCUS

This session explored three key areas that address both U.S. and international hot topics. The first area focused on what states can and should do to prepare their regulatory and legal frameworks for AVs, along with the NHTSA efforts to assist states. A second topic area was the reaction to NHTSA's response to Google concerning the definition of "driver" in the context of AVs. The third area was balancing security, privacy, and innovation in AV data use and ownership.

SESSION SUMMARY

What Can Road Authorities Do to Prepare for Automated Vehicles?

Panelists: Sarah Hunter, Head of Policy, X; Paul Hemmersbaugh, Chief Counsel, NHTSA; David Strickland, Partner, Venable, LLP; and Tilly Chang, Executive Director, San Francisco County Transportation Authority.

What Can Road Authorities, Public Agencies, Municipalities, and States Do to Prepare For Automated Driving?

- Hemmersbaugh suggested that states and cities focus on their traditional areas of authority and regulation. While there may be some gray areas, a lot is black and white in what states and localities have control over. Examples cited were licensing drivers, promulgating and enforcing traffic regulations, and safety inspections. These responsibilities might apply differently to vehicles with varying degrees of AV. A lot of the regulatory questions come down to responsibility—between the federal and state governments—for the regulation, safe operation, and testing and certification of AVs. It would be fruitful to examine the traditional areas of responsibility and review them for possible impediments to AV. It would be desirable to clearly define state and federal spheres where there is clear authority. Considering what might be done differently to promote AVs and how to facilitate use of vehicles with varying degrees of automation would be beneficial. One gray area may be Level 3 operations.
- Hunter suggested that state and local personnel should not panic. The widespread use of AVs is still way off in the future. There is no need to rush to pass laws on different aspects of AVs. She noted that states and cities have asked what they need to do. She suggested that they do not have to do anything. Most laws are currently fit for purpose. Rather than focusing on new laws, she suggested that focusing on what AVs could do for communities would be beneficial. She further suggested getting and staying informed.
- Strickland noted that there are a number of things that governments can do that would be helpful for developing AVs, as well as for traffic safety in general. Examples included enhancing crash and safety data collection, developing active safety systems, and examining issues associated with Level 3 and Level 4 automation.
- Chang discussed issues and approaches in San Francisco. She noted that it is important to focus on the larger topic of how local governments should prepare for any new disruptive service or technology. One takeaway for other cities is to know your policy framework and goals. How new technologies and services can help achieve those goals can then be explored. For example, San Francisco examines new technologies and services through a lens focusing on Vision Zero goals, transit-first policies, and other priorities. This approach helps the public and market understand the city's priorities. Engaging the public and industry partners is important but not always easy. Maintaining an open and inclusive dialog on all topics is important, including TNCs and employer shuttles.

Is There Something Unique About AV That Justifies a Fresh Look? Is There Something That We're Missing That Could Challenge Our Conceptions?

- Hunter suggested that the AV concept is still new and evolving. Maintaining flexibility to accommodate new development is needed. As an example, she noted that no one realized that one unintended consequence of EVs is how quiet they are. Silence can be dangerous for VRUs.
- Chang discussed some of the current challenges associated with employer shuttle bus operations with neighborhoods and curb management policies, which is a local government responsibility. Examining the potential for public–private partnerships is another possible tool for project development and delivery. Considering the types of pilots that are acceptable to a community is also important. She mentioned a project on Treasure Island that is developing a

car-light neighborhood. Involving the community in understanding the ways in which mobility, safety, and other goals can be met is a key element of the project.

What Should Governments Not Do?

- Hemmersbaugh suggested that it was important not to drive hard and fast stakes into the ground on regulatory structures. He noted that flexibility is needed to allow for learning from the deployment of AV technologies, pilots, and demonstrations. He suggested that there is some tension between wanting regulations and guidance to be flexible, adaptable, and iterative to evolve along with the technology, while many companies are saying that more certainty about what the NHTSA and other regulatory environments will look like is needed. One idea would be to begin with a few broad goals and questions that manufacturers' should answer. This approach at the national and state levels would allow for interface between agencies and industry. He noted that the existing regulations have equipped us well to deal with AVs at the state and local levels. But, as adjustments are made for the unique challenges of AVs, regulating in a way that is instructive but also adaptable is needed.

Can You Talk About Tensions and Struggles That Arise When Developing AV Guidance?

- Hemmersbaugh noted that one approach is clearly defining a federal sphere and state and local spheres. While this approach sounds simple, there are many instances where states have a strong legitimate interest in things that NHTSA would like to be principle regulator of, but that it will take time to develop. He suggested that four sections of the anticipated report might be of interest. These sections address operational deployment guidance, model state policy, NHTSA tools and how to tailor them to meet new challenges, and potential new regulatory tools. He suggested the need to avoid a 50-state patchwork of laws, which creates problems for widespread deployment of AV technology.

What Are Your Thoughts About the Suggestion That the United States Should Move to a Pre-Approval Model? Is This a Path Worth Venturing Down?

- Strickland noted that this approach helps create regulatory stability, which is something OEMs are looking for, but that it has limitations related to post-approval. The European and U.S. approaches are different. There are advantages to both, but thoughtful input is needed concerning the best approach to ensure safe operations.
- Hemmersbaugh noted that regardless of whether there is a continuation of the certification system or pre-market approval, or some hybrid, research and development standards are still key. While some think the pre-market would result in faster development, that approach may not be the case.
- Strickland discussed possible staff roles in the post-deployment environment.

Sympathetic to Waiting For Technology to Evolve Before Regulating, But Can We Look Back and See Missed Opportunities to Regulate Technology Earlier or Take a Different Path in the Road (e.g., Loss of Roads as a Public Space During the Introduction of Automobiles)?

- Chang noted that cities are well served by being protective of their vision, rather than being in a reactive mode. San Francisco is well equipped to handle many questions around the potential explosive VMT scenario. There is a need to be ready for tough policy discussions and to deal with trade-offs with long-term goals in mind.
- Panel members and participants discussed data privacy issues and the importance of community involvement and education, public acceptance, and outreach.
- Panel members and participants discussed possible short-term activities, including reviewing existing state and local laws, reviewing the NHTSA guidance when it is released, and continuing to share experiences and lessons learned.

What—Not Who—Are You Licensing as a "Driver"?

James Anderson, *Moderator*

Panelists: Maxime Flament, Department Head of Connectivity and Automation, ERTICO–ITS; Paul Hemmersbaugh, Chief Counsel, NHTSA; Marcus Burke, Project Director for Compliance and Technology, National Transport Commission of Australia; Shin Morishita, Ministry of Posts and Telecommunications; and Geert Pater, Advisor IT Governance, RDW.

Overview of NHTSA Interpretation of Google's Self-Driving System as a "Driver" Within Limited Context

- Burke described the National Transport Commission (NTC) of Australia. The NTC is a policy reform body, not a regulatory agency. It is examining similar challenges to those in the United States in terms of gaining national consistency and working across modes to improve safety, productivity, and environmental outcomes. The NTC was asked last year to review regulatory barriers to automation, including what is required for demonstrations and trials and what is required for AD that requires a driver and AD that does not require a driver. Some of the challenges include national consistency, creating performance-based regulations, managing transition issues regarding mixed fleets, and timing with regulations not moving too quickly or waiting too long.
- Burke discussed the NTC discussion paper released in February for public consultation, with a follow-up released in May with reform options for key issues, including timing of reforms. Near-term issues include supporting trials and clarifying vehicle control. Midterm issues focus on safety assurance, definition of a driver, and enforcement issues. Long-term issues are addressing vehicle standards. Issues that require clarification and refinement include vehicle modifications and maintenance, liability, and privacy.
- Burke reported that the term "automated driving system entity" is used to define what is in control of the car in high automation scenarios. The legal responsibility needs to sit with the entity that controls vehicle. He described the role of government. One question raised through this effort was if governments should simply remove barriers or should governments manage the

technology deployment? Australia currently has a vehicle certification and driver licensing process for human drivers. Who or what is licensed for AD, or if there is a need to license AD systems were examined.

- Burked noted that the NTC reviewed laws for barriers and found 716 provisions that are potential barriers to moving AVs forward. In many current laws, the driver can only be interpreted as meaning a human driver. Driver obligations include safe driving, obeying rules, licenses and permits that must be carried, interaction with enforcement officers, interaction with other road users (providing assistance to drivers involved in a crash), passenger compliance (ensure children wearing seatbelts), provisions not relevant to automated systems (alcohol, drugs fatigue, etc.), and vehicle obligations (roadworthiness, cargo safety, etc.). Two options were examined. The first was to continue the current approach and rely on exemptions. The second was to expand the meaning of driver in relevant legislation to include AD systems. He noted that the NTC prefers the second option. The AD system entity could be any of the relevant companies involved in AV development or operation, such as a manufacturer, car share company, taxi, or freight company. The entity needs the capacity to manage safety risks, have a corporate presence in the jurisdiction, and have appropriate insurance.
- Burke provided two options for a safety assurance framework for driverless vehicles. The first option focused on removing barriers for high-level AVs to enter the market, and allowing the industry to self-regulate. The second option would implement a national safety assurance framework to oversee deployment of high-level AVs. He noted that the NTC supports the second option for the medium-term. Elements in the safety assurance framework might include the safety management system, risk management, data sharing and incident notification, security, and a clear definition of scope of operations. It could be similar to what is currently used in rail and aviation sectors. It would provide flexibility, leaving it to industry to identify the mix of technologies. He noted that recommendations are to be made in November 2016.
- Flament noted the challenge of presenting the EU perspective since each member state has the ability to regulate independently. In consulting with other member states for the presentation, all refer to the 1968 Vienna Convention on Road Traffic. The Vienna Convention provides a definition of driver as "any person who drives a motor vehicle or who drives animals on the road". He described the Volvo DriveMe example where people are told that they can detach from the primary task of driving. Volvo has accepted full liability when the vehicle is in the autonomous mode. Volvo is limiting risk by limiting the geographic extent of operation and conditions in which the autonomous mode is engaged. The "driver" is the entity who accepts the liability for a crash. In this case, the driver is Volvo. A second example is an AV shuttle with no manual controls aside from an emergency stop button. The intention is for the vehicles to be operated remotely. The definitions in revisions to the Vienna Convention are as broad as possible. Road vehicles may take many shapes and the Convention must apply to all of them. The driver may be inside or outside the vehicle. Control can take many forms—indirect or direct. The driver has the ultimate responsibility to operate the vehicle, but may not be behind the steering wheel and may not be in direct control. The identification of the driver is important to determine who is liable for actions of the vehicle. Control may be delegated, but the driver is responsible.
- Flament noted that future work focuses on clarifying issues related to driver injury insurancc, defining data recording regulation and data access rights, and clarifying liability in case of poor maintenance, inappropriate use, hacking, and OTA updates.

- Morishita noted that Japan is in an early stage of evaluating the legal framework for automated vehicles. He described the ITS promotion framework of the Japanese government. The National Police Agency has enforcement over road traffic safety. The Ministry of Land, Infrastructure, and Tourism has responsibility for developing and enforcing safety regulations for road vehicles and onboard equipment. These two entitles will be asked to find solutions for AV legal issues. He discussed that it is important to verify functionality and safety of each technology for AD systems for field operational tests on public roads. The Counsel for Science, Technology, and Innovation coordinates implementation of AD systems, and promote research and development on AV driving systems.
- Morishita discussed the SIP-adus program, which was launched in 2013. The first priority of SIP-adus is achieving the national goal reducing road crashes. The annual target is to reduce fatalities by 2,500 per year through 2020. To reach that goal, SIP-adus is moving forward on AV research and development. AV driving systems will be realized through integrating onboard technologies. The interface and transition between computers and humans is also important. Even in Level 4, there is a need to consider interactions between self-driving vehicles and other road users. International cooperation is critical to addressing challenges.
- Morishita noted that the National Police Agency has developed guidelines for public road testing of AV systems. Testing must comply with other regulations. An individual must be in driver's seat and must operate a vehicle in emergency situations when necessary. Vehicles must be tested at test facility beforehand. The guidelines state that the system should be able to request that the driver take over operation in advance, so there can be a smooth transition. Unless it is a full self-driving system, this point of view should be considered in discussing the relationship between AVs and human drivers. SIP-adus has set five focus areas for field operation tests: dynamic map, HMI, cybersecurity, pedestrian assistance, and next-generation public transit. It is intended that field operation tests will be open to domestic and international experts.
- Pater noted that the RDW is the Dutch vehicle authority. It conducts and is responsible for approvals with other EU countries for licensing, license plates, and issuing driver's licenses. He said that exemptions for OEMs to operate AVs on Dutch roads may be given on the condition that they bear liability. Every law says there has to be one liable person/entity in charge of the vehicle. He noted that there is a lot to investigate regarding AVs. He suggested that the driver will stay the driver, such as the human behind the steering wheel, until Level 5 is reached.
- Hemmersbaugh reviewed the Google request to NHTSA on the interpretation of existing regulations. The NHTSA regulations define "driver" as the occupant of motor vehicle seated immediately behind steering control system. The Google vehicle does not have a steering wheel, steering column, gas pedal, or brake. The NHTSA regulations assume traditional vehicle design, not what Google is developing. Many regulations or testing procedures reference the steering wheel or driver. NHTSA concluded that Google's self-driving system functioned as the driver in the context of NHTSA regulations regarding Google's vehicle. He noted that it was not clear what future direction NHTSA would take. Many requirements do not make sense with a software system as the driver. Under NHTSA's view of the law, interpretations are simply explanations of how regulations apply in a new context.

APPENDIX C: BREAKOUT SESSION 3

Human Factors in Road Vehicle Automation

Organizers

- Daniel V. McGehee, University of Iowa;
- Ashley McDonald, University of Iowa;
- Alex Epstein, National Safety Council;
- David Yang, FHWA; and
- Brian Philips, FHWA.

Reporters

- Ashley McDonald, University of Iowa;
- Brian Philips, FHWA; and
- Katherine Turnbull, TTI.

SESSION FOCUS

The human factors group sponsored two sessions. The first session focused on preparing the public for AVs and the changing transportation network. The session goals were to bring together key stakeholders involved in driving and technology consumer education, to identify educational needs associated with AVs, and to identify challenges to future consumer education efforts. The second session examined human factors issues associated with near-term deployment of connected and automated applications, including lateral and longitudinal control, such as CACC. Discussion focused on ensuring safe operations on freeways and interactions with other vehicles as these systems are deployed, as well as needed infrastructure elements for safe operations. The PowerPoint presentations used by some of the speakers are available on the AUVSI website at http://www.automatedvehiclessymposium.org/home.

SESSION SUMMARIES

Session 1: Look Ma', No Hands! MyCarDoesWhat? Preparing the American Public for AVs and the Changing Transportation Network

Advanced Technology Integration: A Bird's Eye View of Where We've Been and Some Observations Toward the Future

Linda Angell, *Touchstone Evaluations, Inc.*

This presentation provided a summary of the development, introduction, and use of advanced automotive technologies over the past two decades. Angell noted that the experience highlights that technology development and integration as a process is both difficult and multifaceted. She also discussed the consumer education components associated with the different technologies, which tended to occur at the end of the process and which were also difficult and complicated.

Bernard Soriano, *California Department of Motor Vehicles*

This presentation focused on the California experience with consumer education and outreach related to developing requirements for testing autonomous vehicles in the state. Soriano discussed the challenges identified by the state, including developing a minimal set of competencies. He noted that it is not possible to test for every scenario. He discussed elements for other states to consider with the deployment of automated vehicles, noting that every state has a unique set of considerations. The NHTSA model state policies and guidelines is an excellent start, but differences between jurisdictions need to be recognized and jurisdictions need to have the flexibility to craft guidance to meet their situations. He also suggested that testing needs to evolve as technology evolves.

Daniel McGehee, Ashley McDonald, and Alex Epstein, *University of Iowa; MyCarDoesWhat.org*

This presentation described the national campaign MyCarDoesWhat.org, which focuses on educating drivers about ADAS. The outcome-driven approach includes a website with videos, course materials, and interactive games. The use of these features was highlighted, along with future activities. The speakers noted that many drivers do not know the current technology on their vehicles.

Omar Ahmad, *National Advanced Driving Simulator, University of Iowa*

This presentation focused on the development and capabilities of a virtual driving world for use in testing human drivers interacting with partial and complete CAV systems. Examples of the use of the driving simulator were highlighted.

Oliver Bayley, *Peloton Technology, Inc.*

This presentation described public driver misperceptions associated with truck platooning, challenges to driver acceptance of platooning, and potential benefits from platoons. The approach Peloton is using to develop a truck platooning system was discussed.

SESSION DISCUSSION

Session participants discussed the need for public education and outreach to support the introduction and use of automated, connected, and autonomous vehicles. Different methods were identified for providing information on the technologies available on vehicles and how they work. Providing information on technologies on other vehicles, such as truck platooning, and on the infrastructure, was also discussed by participants.

Session 2: Human Factors Issues Associated with Connected Automation Applications

A panel on Human Factors Findings from a Connected Automation Application—Cooperative Adaptive Cruise Controls

David Yang, *FHWA, moderator*

- Opportunities for Connected Automation Applications to Improve Mobility and Energy Use. Robert Ferlis, FHWA.
- Findings from CACC Experiments: Overview and Merging. Brian Philips, FHWA.
- Findings from CACC Experiments: Car Following and Collision Avoidance. Vaughan Inman, Leidos.

The three speakers discussed CACC research conducted at the Turner–Fairbanks Highway Research Center. They described a CACC simulation study, CACC physical performance testing, automated lane changing, and speed harmonization research. Research focusing on testing an eco-driving approach and departure at a signalized intersection was also highlighted.

Small Group Discussions

Session attendees were divided into small groups and asked to consider key human factors questions related to connected automation that need to be answered when:

- Drivers have information on the surrounding pedestrians, bicyclists, and critical elements of the infrastructure that is provided by the advances in connectivity; or
- Drivers have information on the other vehicles near their own vehicle that is provided by the advances in connectivity.

Facilitated Expert Panel Discussion

Don Fisher, *Volpe National Transportation Systems Center, facilitator*

Expert Panel Members:

- Jessica Cicchino, Insurance Institute for Highway Safety;
- Charles Green, General Motors;
- John Lee, University of Wisconsin;
- Michael Regan, Australian Road Research Board (ARRB);
- James Sayer, University of Michigan Transportation Research Institute; and
- Heishiro Toyoda, Toyota Technical Center.

Panelists discussed some of the major human factors issues that need to be addressed with the design, development, and deployment of connected automation applications. Methods to integrate consumers into the design and development process for automated, connected, and autonomous vehicles were discussed by panelists and participants. Possible ways to bring together OEMs, software developers, government agencies, academia, and marketing were also discussed by participants.

MOVING FORWARD

- Participants discussed the importance of continuing human factors research associated with a range of topics related to AVs, CVs, and autonomous vehicles.
- Participants discussed holding a human factors breakout session at the 2017 AVS.

APPENDIX D: BREAKOUT SESSION 4

Impact Assessment

Organizers

- Scott Smith, Volpe Center, U.S. DOT;
- Satu Innamaa, VTT Technical Research Centre of Finland;
- Ching-Yao Chan, California PATH/University of California, Berkeley;
- Bo Ekman, Danish Road Directorate;
- Natalia De Estevan-Ubeda, Transport for London;
- Steve Lockwood, Steve Lockwood, LLC;
- Richard Mudge, Compass Transportation & Technology;
- Yves Page, Renault;
- Lydia Rainville, Volpe Center, U.S. DOT;
- Nick Reed, Transport Research Laboratory;
- Simon Tong, Transport Research Laboratory;
- Isabel Wilmink, TNO (Netherlands); and
- Chris Williges, HDR.

Reporters

- Scott Smith, Volpe Center, U.S. DOT; and
- Satu Innamaa, VTT Technical Research Centre of Finland.

SESSION FOCUS

The purpose of this breakout session was to discuss the assessment of direct and indirect impacts related to AVs. This was accomplished by applying systems thinking to the impacts of AVs (Figure 1). One goal of the session was to identify ways to improve the design of field operational tests to maximize the insights obtained. Improving the design of field operational tests includes finding appropriate performance indicators and establishing a baseline. It is also essential to identify the most important direct and indirect impacts and the main linkages between them. The presentations in this session aimed to stimulate discussion about all of these matters. Information on these impacts and their outcomes will enable decision makers and researchers to address the investment and policy decisions needed today to make desired outcomes more likely. PowerPoint presentations used by some of the speakers are available on the AUVSI website at http://www.automatedvehiclessymposium.org/home.

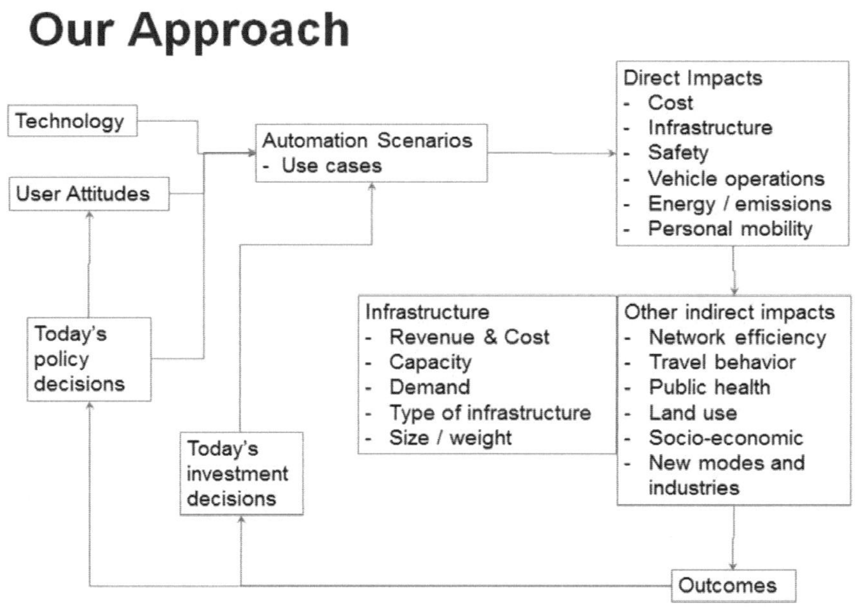

FIGURE 1 Impact assessment approach.

SESSION SUMMARIES

Welcome

Scott Smith, *Volpe Center, U.S. Department of Transportation*

Smith provided a brief introduction of the session and reviewed the purpose and goals of the breakout session. Examples of impact linkages were introduced (Figure 2). These included the impacts of AVs on land use, personal mobility, public health, travel behavior, and network efficiency.

Direct Impacts: Introduction and Case Studies

Introduction

Satu Innamaa, *VTT*

Direct impacts include cost, infrastructure needs, safety, vehicle operations, energy, emissions, and personal mobility. These impacts can usually be measured by field operational tests. Relevant questions in measuring direct impacts are:

1. What are appropriate performance indicators?
2. How does one establish a baseline for comparison?
3. How does one design an AV field test or simulation to gain the appropriate insights?

and

4. What data can be shared with third parties?

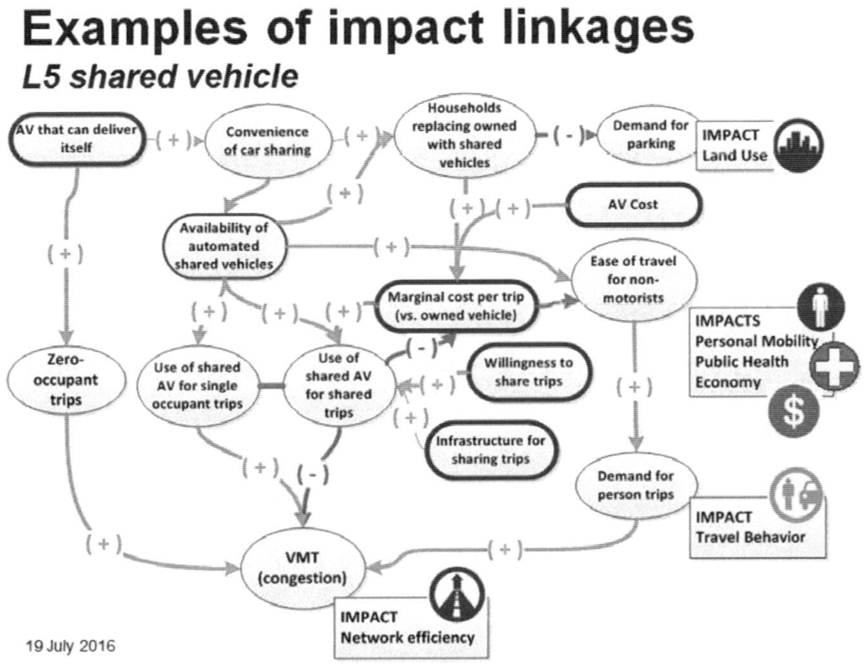

FIGURE 2 Examples of automated vehicles' impact linkages.

Direct impacts will lead to indirect impacts, and the distinction between direct and indirect impacts is not always simple.

Automated Low-Speed Shared Shuttle

Nick Reed, *Transport Research Laboratory, U.K.*

Automated low-speed shuttle buses travel on a set of routes within a city region. These vehicles are designed for heavy utilization. The vehicle's speed is 15 mph in pedestrian areas. Direct impacts may include low cost, safety benefits, limited infrastructure requirements, good accessibility (for example serving individuals in wheelchairs), and reduction in emissions. There is also a vision that people would walk and bicycle more since the environment is less noisy around automated shuttles.

Truck Platooning

Isabel Wilmink, *TNO, Netherlands*

Platooning trucks have SAE Level 4 automation. Platoons can be formed in real time. Expected direct impacts include reduced costs, need for physical and virtual infrastructure, increases in safety, more stable traffic flow, reduced fuel consumption, and reduced emissions. Personal mobility may be affected if the truck drivers are able to engage in nondriving activities, potentially leading to more freight miles since the cost is lower.

Midlevel Automated Passenger Car (Drive Me)

Trent Victor, *Volvo*

With these vehicles, the driver has an opportunity to do something else while driving. However, the Level 4 automation system offered by Volvo may not work in bad weather conditions and in tunnels. Traveling is estimated to be safer. Precautionary drivers learn from risky situations and strive towards safer driving. Product development of these vehicles includes safety verification process and safety impact analysis.

Private Fully Automated Vehicle

Scott Smith, *Volpe Center, U.S. Department of Transportation*

Personal AVs shared only within a household and shared AVs may have very different impacts on parking requirements and demand for travel.

AUDIENCE DISCUSSION

The participants were asked to divide into groups and select one of the cases for closer consideration. Groups were asked to identify the most important direct impacts and appropriate performance indicators, and to design an AV field test or simulation to gain the appropriate insights. Finally, they were asked to discuss which data could be shared with third parties. At the end of group discussion session, each group reported back on their results and insights.

Low-Speed Shuttle

The group discussing automated low-speed shared shuttles presented the following performance measures and questions:

- Amount and location of land required for parking.
- Mix of people being served with the shuttles.
- How shuttle rides traveled before the mode was available. For example, will the number of pedestrians decrease since the shuttle is a low-cost option?
- Propulsion, energy, and emissions for the shuttle.
- Use attitudes: Do the passengers feel safe without a driver?
- Other road user attitudes: Will other road users feel frustrated because the shuttle's speed is so low?

Participants noted that an actual shuttle project, GATEway, will address some of these questions via a two-phase trial. The first phase includes an in-depth interview of different demographic groups in order to gain understanding about their needs and concerns. The second phase would focus on doing more information testing with larger groups.

Truck Platooning

Participants identified several performance indicators:

- Safety, including surrogate measures such as the number of near-crash situations and changes in behavior of other vehicles around the platoons;
- Fuel consumption;
- Vehicle utilization or the share of time that the truck is in use;
- Driver productivity, including the share of time spent driving, working but not driving, resting;
- Need for truck parking areas, which may be reduced if there is less need for trucks to stop;
- Type of trucks and what they carrying; and
- A baseline would be existing trucking operations. A two-stage field test may be appropriate. The first would focus on vehicle operations (fuel consumption, safety), working with vehicle manufacturers. The second would focus on user issues, working with fleet owners.

Midlevel Automated Passenger Car

The midlevel automated passenger car case stimulated a discussion about the effect of incidents on congestion and characterizing impacts as direct or indirect. The possible increase in safety leads to reduced crashes, which in turn leads to reduced congestion, since congestion is often caused by accidents. On the other hand, AVs may affect congestion in other ways as well (e.g., through changed following distances), so it may be difficult to distinguish direct and indirect impacts. Identified performance measures besides safety and congestion are changes in driver behavior, human vehicle interaction, and interaction with other vehicles.

Fully Automated Passenger Car

The discussion on fully automated passenger cars resulted in a number of observations. First, it is essential to understand the use case, technological and political capabilities, and time frame for an impact assessment. Second, the economics of how these vehicles are developed is important, as the cost of the vehicles will affect ownership and adoption rates. Questions about ownership and the number of vehicles per household are also important. Third, the degree of municipal funding will characterize the design of the system.

Indirect Impacts: Introduction and Case Studies

Introduction

Steve Lockwood, *Steve Lockwood, LLC*

Direct impacts are unmediated and immediate in time and place. Indirect impacts, in turn, are mediated by other factors and removed in time and place. Direct impacts can lead to multiple primary and secondary indirect impacts.

Indirect impacts concern network efficiency, asset management, public health, land use, socioeconomic impacts, and others. Reduced need for new capacity has an impact on transportation system maintenance and operations. Reduced land for parking may enable complete streets, with improved walking and bicycling accommodations. Increased safety can lead to a reduced need for insurance. Better air quality may increase public health. Better transportation may improve access to employment.

There are many research issues that closely relate to the indirect impacts of AVs. For example, how to model and forecast with multiple mediating factors and major uncertainties. The uncertainty level is high when studying indirect impacts and predictions are highly dependent on the assumptions that are made.

Urban Shared Shuttle

Nick Reed, *Transport Research Laboratory, U.K.*
Scott Smith, *Volpe Center, U.S. Department of Transportation*

Case study 1 focused on an urban shared shuttle. Reed introduced the benefits of the shuttles and Smith introduced some of the related challenges and constraints. A number of questions arose. How will the vehicles be powered? How comfortable are the passengers going to be? How do the shuttles affect urban infrastructure? How would the payment be arranged?

Truck Platooning

Isabel Wilmink, *TNO, Netherlands*
Bo Ekman, *Volpe Center, U.S. Department of Transportation*

Case study 2 focused on truck platooning. Wilmink introduced the benefits of truck platooning and Ekman described the related challenges and constraints. The European truck challenge provided extensive lessons and opportunity to learn. He noted that that the promising energy reductions may be hard to achieve in practice. One of the key questions here is what it takes to get to get enough trucks willing and able to platoon to make it work? Truck platooning on a larger scale needs fleet managers and drivers' willingness to cooperate.

The presentations and discussions in this breakout session led to a number of insights and lessons learned. First, impact mechanisms are complex and far-reaching. The impact mechanisms include interactions between direct impacts and indirect impacts, and they vary from short-term impacts to very long-term ones. It is necessary to keep in mind that most important impacts are different for different people—a positive impact for one can be negative for someone else. Another essential thing is to clearly define the use cases and context. This means defining the environment, the time scale, perception, and other parameters. It is a challenge to consider the future uncertainty in today's policy and infrastructure decisions. To address these uncertainties, ranges should be used, rather than single values.

MOVING FORWARD

- Based on this breakout session on impact assessment, research issues were identified. First, it is important to acknowledge the significant uncertainty in automation. The cases that were envisioned will become different in reality. The need to incorporate uncertainty explicitly, and design for flexibility was suggested. The definition of what is meant by AV is often unclear (is the self-driving car at SAE Level 4 or Level 5? There is a big difference). There is a large range in the projections for when AVs will arrive. Gartner's hype cycle was mentioned. It is important to manage the expectations of the public. Finally, will technology arrive faster than policy? For example, is DSRC the only way to achieve the benefits of connected vehicles?
- Participants noted that 100% penetration rates will not be achieved immediately. Research is needed on the impacts of different applications at different penetration rates. At some level of market penetration, it may become attractive to modify infrastructure or policies (e.g., the establishment of AV-only managed lanes).
- There is a wide range in vehicle types, in terms of size, weight, and propulsion. Participants noted the need to distinguish between passenger and goods movements. Propulsion (electrification) is an open question. For example can dual-mode (electric, internal combustion) trucks be geofenced to run in zero-emissions mode in sensitive areas. Should there be investments in dynamic wireless charging on Interstates and motorways, so that vehicles do not need big batteries?
- Participants discussed the importance of identifying the benefits and costs of these projects. How do these benefits accrue to different parts of society? Identify the benefits that require public sector interaction. Is public sector funding available?
- Participants noted that it was important to provide enough detail in publications to enable readers to assess relevance. This includes defining what is meant by "automation," as well as clearly defining the use cases and context. The automobile insurance sector is already using telematics to identify low-risk drivers. Might the insurance industry validate some of the research? One participant noted that agent-based modeling is being used to model energy impacts. Some data are available now, and can be used to test the reasonableness of the assumptions. For example, if the amount of carpooling in the United States has been declining, is it realistic to assume that most users will be willing to share trips? Additionally, in urban areas, the search for parking has been said to account for a significant portion of VMT. Is this true and what will be the impacts of automation?
- Participants discussed that it is still important to better identify high-level indirect impact mechanisms for automation. Unintended impacts should be studied (an example mentioned in the plenary was that of reduced crashes reducing the supply of organ donors). Finally, it was suggested that a commonly agreed-upon impact assessment framework could facilitate the sharing of insights.

APPENDIX E: BREAKOUT SESSION 5

Enabling Technologies

Organizers

- Jim Misener, Qualcomm Technologies, Inc.;
- Cristofer Englund, Viktoria Swedish ICT;
- John Estrada, eTrans Systems;
- Juhani Jaaskelianen, Consultant;
- Frank Serna, Strategic Initiatives, Draper Laboratory;
- Surya Satyavolu, Sirab Techonologies; and
- Sudharson Sundararajan, Booz Allen Hamilton.

Reporters

- Jim Misener, Qualcomm Technologies, Inc.; and
- Katherine Turnbull, Texas A&M Transportation Institute.

SESSION FOCUS

One goal of this session was to explore a wide range of technologies needed to establish AVs including position, localization and mapping; algorithms, deep learning techniques, sensor fusion, guidance and control; hybrid communications; sensing and perception; and data ownership and privacy. A second goal was to gain an understanding of how these technologies will work together to address the needs of the various applications, with recognition of data ownership, regulatory, and standardization perspectives. A third goal was to realize the potential shortfalls in these technologies. The PowerPoint presentations used by some of the speakers are available on the AUVSI website at http://www.automatedvehiclessymposium.org/home.

SESSION SUMMARY

The session included five panels, with speakers providing brief comments. The topics addressed by panelists providing their PowerPoint presentations are highlighted.

Position, Localization, and Mapping

Mark Tabb, *HERE*
Sravan Puttagunt, *Civil Maps*
Rob Hranac, *Swift Navigation*

- Tabb described the HERE Automated Driving Framework, which includes HDMap, Live Roads, and Humanized Driving. HDMap is the foundational component, providing a detailed, precise lane-level model that includes a localization object layer.

Algorithms, Deep Learning Techniques, Sensor Fusion, Guidance, and Control

Trevor Darrell, *PATH*
Xinzhou Wu, *Qualcomm Technologies, Inc.*
Shahrokh Daijavad, *IBM*

- Daijavad discussed IBM products and analytic capabilities, including using IBM Watson IoT for an automotive platform. Watson-on-Wheels provides innovative on-vehicle analytics for driver behavior, road mapping, and fleet management.
- Wu discussed some of the challenges associated with enabling mass adoption of autonomous driving. Examples included the need for low-cost sensors with complex algorithms and low-power computing platforms with advanced algorithms and hardware design.

Hybrid Communications

Gaurav Bansal, *Toyota InfoTechnology Center, USA*
Thierry E. Klein, *Nokia*
Sanjeev Athalye, *Qualcomm Technologies, Inc.*

- Bansal discussed hybrid communications to improve safety. He described the advantages, limitations, and uses of different technologies in AVs, and presented a proposed system model. His key take aways were that V2X communication will play an important role in reducing crashes and that high bandwidth, low-latency communications is required for AD.
- Klein described communication technologies for AVs with a focus on cellular technologies. He outlined V2V–infrastructure communications via LTE and LTE-X. He described the benefits of cellular communication technologies and presented potential applications.
- Athalye discussed V2X as an enabler of autonomous driving. He highlighted the vision for the always-connected vehicle of the future as highly secure, highly intelligent, always connected, increasingly autonomous, and increasingly electric or hybrid. Accomplishing this vision requires new levels of connectivity and intelligence. V2X enables a broad set of use cases and is a key to ADAS. He also noted that V2X requires a regionally harmonized ITS spectrum, with at least 70 MHz of spectrum recommended.

Sensing and Perception

Roger Berg, *DENSO International America, Inc.*
Michael Maile, *Daimler AG*

- Maile highlighted vehicle automotive technology. He described the typical components, including sensors on vehicles, the actuators, the operating system, and the back office support system.

Security Innovation, Technologies for Data Ownership and Privacy

Angelos Amditis, *Institute of Communication and Computer Systems*
Jonathan Petit, *Security Innovation, Inc.*

- Amditis discussed open access of data and data protection and privacy. He described factors influencing the need for an open platform and possible platform concepts, including an onboard application platform, in-vehicle interface, and data server platform. He outlined possible threats and risks to data privacy. Amditis noted that the OECD principles and general data protection regulation are reflected in the European Directive 95/46/EC. He reviewed the privacy principles and privacy policies, and legal and technical approaches.
- Petit discussed data ownership and privacy. He described different options of vehicle and data ownership. He suggested that factors to consider in data ownership include the type of data, who has access to the data, what the data is used for, what the user gains for giving data away, and the length of time the data is available. Groups interested in AV data include OEMs, suppliers, TMCs, law enforcement agencies, insurance companies, data brokers, advertising companies, researchers, fleet managers, and users. Technologies are available to protect data privacy.

OPENING DISCUSSION

- Participants discussed opportunities for synergy related to digital mapping systems, positioning, and the infrastructure through the use of crowdsourcing and road features. Another opportunity for synergy focused on hybrid communications of difficult use cases, such as collaborative long- and short-range.
- Speakers and participants discussed that the perspectives of deep learning may enable the use of low-cost sensors and provide the opportunity to leverage available computational resources.
- Participants and speakers discussed data ownership and privacy issues, and noted that these concerns are as essential as engineering and policy considerations.

MOVING FORWARD

- Participants discussed that developing a summary of contemporary research directions and would be beneficial.
- Participants also favored continuing the discussion at the 2017 AVS, including providing more time to explore and catalog synergisms between enabling technologies in more detail.

APPENDIX F: BREAKOUT SESSION 6

Safety Assurance

Organizers and Reporters

- Herman Winner, Technische Universität Darmstadt, Germany; and
- Ching-Yao Chan, California PATH Program, University of California, Berkeley.

SESSION FOCUS

With the rapid progress in functional skills of AD, the question of how the safety of AVs can be assured increasingly becomes key to the introduction of AVs to public operations in real world environments. Until now, there have been no well-defined standards or commonly accepted procedures to validate the safety of AVs.

In this breakout session, different approaches to safety assurance from a variety of projects preparing for the introduction of AVs were presented. The session also includes presentations and facilitates discussions of the requirements on safety assurance from societal perspectives and on the feasibility of satisfying such requirements within the current or expected technological constraints.

GOALS AND OUTPUTS

- Presentations by invited speakers delivered an overview of the safety assurance approaches and respective methods.
- Participants obtain an understanding of the discussed methods and engage in discussions to convey different perspectives.
- Panel discussions identified the next steps for forming a strategy for achieving the goals and objectives of safety assurance.

SESSION SUMMARY

Nine presentations were given across two sub-sessions. The first session included four speakers followed by a panel discussion, and the second subsession consisted of five presenters followed by a panel discussion as well. PowerPoint presentations used by some of the speakers are available on the AUVSI website at http://www.automatedvehiclessymposium.org/home.

SUBSESSION 1

Speakers	Summary of Presentation
Jonas Nilsson, Volvo Car Corporation, Gothenburg	**Dependability and Verification for Self-Driving Cars: The Drive Me Approach** The development of self-driving cars represents a paradigm shift for the automotive industry. This talk was based on experiences from the Drive Me project in Gothenburg, Sweden, and focused on the new challenges in safety and dependability brought by autonomy. In addition, the impact these challenges have on safety assurance and verification were elaborated.
Naohisa Hashimoto, National Institute of Advanced Industrial Science and Technology (AIST), Japan	**Concerning Safety Assurance on Automated Vehicle: Results and Discussion Based on the Projects in Japan** The safety of automated vehicles depends on several aspects including sensors, algorithms, system architectures, and human factors. Different automated vehicle projects can be categorized according to the objectives, and safety should be evaluated considering each category. This was discussed with lessons learned from the automated vehicle projects in Japan.
Walther Wachenfeld, Technische Universität Darmstadt (Germany)	**Safety Assurance Based on an Objective Identification of Scenarios: One Approach of the PEGASUS Project** Assessing automation by test driving is economically not feasible ahead of market introduction. Thus, testing has to be shifted to other testing tools. These tools need information on the relevance of scenarios to reduce the test effort. The key question is what is of relevance when assessing AD? Within the PEGASUS project one primary goal is to answer this question for HAD on motorways. The presentation proposes and discusses an objective identification of scenarios and their relevance for assessing safety.
Lutz Eckstein, RWTH Aachen University (Germany), Institute for Automotive Engineering (ika)	**Developing and Assessing Automated Driving** The assessment and certification of AD probably constitutes the most demanding challenge, which requires a sophisticated, collaborative approach. In this presentation a novel approach to structure and visualization of the interdependencies of challenges was proposed. Special emphasis is put on the question of assessment and certification on the way to automate driving, and the contribution of different projects was described.

SUBSESSION 2

Tim Allan Wheeler, Stanford University, Stanford Intelligent Systems Laboratory	**Establishing Trust in Autonomous Vehicles: An Aerospace Perspective** AVs and other emerging active driving systems require advanced science and engineering methodologies by which trust can be established. The route to building trust lies in the creation of a scientific, unified, transparent framework to optimize and evaluate active driving systems. A cross-industry standard model by which the safety of active driving systems can be evaluated, based on approaches successfully applied in international civil aviation, was proposed.

Nidhi Kalra, RAND Center for Decision Making Under Uncertainty	**Driving Autonomous Vehicles to Safety** The presentation raises the following questions: How safe should AVs be before they are allowed on the roads? How do we (not) prove they are safe? How might our near-term safety choices affect the long-term evolution of the technology? These and other pressing policy questions are discussed. Suggestions are made to explore how adaptive regulations may be a promising way to answer the aforementioned questions.
Marcos Pillado, Applus IDIADA, Spain	**Functional Validation and Performance Assessment of Automated Truck Platoons in Controlled Environments** Platooning of heavy duty vehicles provides the opportunity to save fuel, increase safety, and add road capacity. The cooperative dynamic formation of platoons for safe and energy-optimized goods transportation (COMPANION) project aims to develop and validate a system for creation, coordination, and operation of platoons. A complete integration of the entire system is performed in the project in order to make a global assessment of the full system. The testing methodology used for the validation and performance assessment of the platooning maneuvers and the onboard HMI in a controlled scenario is presented.
Michael Wagner, Carnegie Mellon University	**Challenges in Autonomous Vehicle Testing and Validation** Software testing is all too often simply a bug hunt rather than a well-considered exercise in ensuring quality. A more methodical approach than a simple cycle of system-level test–fail–patch–test will be required to deploy safe AVs at scale. The presenter identified five major challenge areas in testing according to the V model for AVs and discussed promising potential solutions. While significant challenges remain in safety certification of algorithms that provide high-level autonomy themselves, it seems within reach to instead architect the system and its accompanying design process to be able to employ existing software safety approaches.
Andrew Lacher, Unmanned and Autonomous Systems Research Strategist, MITRE Corporation	**Applicability of Lessons Learned from Aviation Safety Management System for Automated Vehicles** In January 2016, U.S. DOT, NHTSA, and the automotive industry agreed to examine the existing aviation industry voluntary/anonymous safety information reporting systems to understand whether such an approach could be utilized in the automobile sector. Safety data sharing is one of the components of a safety management system (SMS), which is a standard recognized throughout the aviation industry worldwide. Using SMS practices, the Commercial Aviation Safety Team (a combined industry/government group) reduced the risk of commercial aviation fatalities in the United States by 83% in 10 years. The presentation also addressed lessons learned from SMS, similarities and differences between the aviation and auto sectors, and how the movement toward automated and connected driving effects of SMS-like practices.

SUMMARY OF PANEL DISCUSSION

- What is the role of the driver for different levels of autonomy? The experts explained the differing expectations but there was still reluctance whether customers would understand the differentiation. The challenges associated with driver hand-off were emphasized.
- How can data for safety assurance be collected commonly and shared for the design of safety systems? The German government has started a project to collect data for certification purposes.
- How can learning systems be tested and why would monitoring unknown situations be useful? There is no solution for the problem of unknown unknowns. One can have a run-time monitor to record a system's boundary and capabilities then the behavior can be pre-identified to ensure the whole system behavior is safe. The monitor can tell whether the driver is in a known condition, rather than a safe condition; or if not, the system can issue a reasonable fail-safe strategy.
- How to communicate the fact that the AV cannot not be shown to be perfect before introduction? There is a fundamental difference between opinions and values. We should clearly explain the implications and have it accessible to the public. It is a communication problem, but mere communication is not enough. How a message is conveyed to the public greatly influences interpretation. There is a perception gap about what the testing is for. It is important to make the communication clear, so people could think through the problem more carefully.
- Is there anything to be learned from the FAA to apply to autonomous driving? The comparisons between the worlds of aviation safety and road safety were discussed concerning risk management and their financial resources for safety. The FAA Aviation Safety Information Analysis and Sharing project could be used as a reference for an automotive data and safety reporting program.

The following is a list of key points captured in the panel presentations:

- A safe- and fail-operational vehicle implies a lot of redundancy, which is not a current state-of-the-art practice in automobiles.
- Safety-related tasks must be clearly divided between driver and autopilot.
- There are strong interdependencies between levels of automation and safety assurance approaches.
- Scenario-based testing needs a documented and traceable way showing where the tests are derived from.
- Databases and test methodologies must be developed and deployed step-by-step.
- A scientific, unified framework to optimize and evaluate the safety will lead to trust in AD.
- We cannot wait for a perfectly safe AD system. Instead, we should begin with reasonable safety in order to offer improvements as soon as possible with the potential benefit of saving lives.
- (Machine learning) AD (sub-) systems should be tested for robustness with the most challenging test cases.
- From aviation we can learn more for safety management processes than from their methodology.

MOVING FORWARD

The following is a list of next steps suggested by the panel participants:

- Creating open databases for test scenarios;
- Establishing a safety management system that could mimic the aviation model;
- Developing scientific accepted validation methodology;
- Establishing standards, for validation, not for function; and
- Convey realistic expectations about safety to the public.

APPENDIX G: BREAKOUT SESSION 7

Future Challenges for Automated Trucks

Organizers

- Daniel Bartz, Automation;
- Karlyn Stanley, RAND;
- Mohammad Poorsartep, TTI;
- Osman Altan, FHWA;
- Richard Bishop, Bishop Consulting;
- Steve Shladover, California PATH Program, University of California, Berkeley;
- Steve Underwood, University of Michigan Transportation Center;
- Thomas Stevens, Argonne National Laboratory; and
- Tom Voege, International Transport Forum–OECD.

Reporter

- Thomas Stevens, Argonne National Laboratory

SESSION FOCUS

This session focused on the challenges and opportunities for deploying various levels of automated trucks. The session consisted of three sections: (1) discussion of truck platooning and truck automation activities and policies in Europe; (2) a panel of diverse stakeholders discussing truck automation; and (3) a discussion of full truck automation. Finally, the group recapped the highlights of the breakout session.

SESSION SUMMARY

The European Truck Platooning Challenge

Tom Alkim of the Rijkswaterstaat of the Netherlands (the Dutch Ministry for Infrastructure and the Environment), gave a presentation on the European Truck Platooning Challenge. This was a demonstration of two- and three-truck platoons (six total, one each from DAF Trucks, Daimler Trucks, Iveco, MAN Truck & Bus, Scania, and Volvo) that drove different routes through several countries (Sweden, Denmark, Germany, and Belgium), all ending up in Rotterdam, Netherlands. The challenge required an agreement between the six truck manufacturers, country and province-level road authorities, labor unions, and the EC. Many legal and regulatory issues had to be addressed by the many different jurisdictions in countries and regions involved. A total of 19 exemptions to various regulations were obtained to address anticipated risks of platooning, including:

- Risk of accidents or disturbance of traffic flow;
- Increase bridge and road wear;
- Complex traffic behavior;
- Transition of truck control (from automated system to the driver); and
- Possible failure of the automation system.

Risk mitigation measures were documented in the exemptions, which are publicly available. These included decoupling platoons at bridges and on-ramps and off-ramps.

Alkim stressed that the Truck Platooning Challenge was not a research project, but a demonstration intended to build support and momentum for truck platooning. However, valuable information was obtained during the challenge, from interviewing drivers and from aerial footage of platoons. Interviews revealed that when platooning, drivers of the lead truck felt responsible for the whole platoon and drivers in a platoon tended to want to keep the platoon together. They found that interactions between platoons and other trucks were more complicated than interactions with cars. This point was evident in one excerpt of the aerial footage that Alkim showed in which a two-truck platoon decoupled as it approached an on-ramp. As the following truck slowed down, another truck behind it changed lanes to pass. At the same time, a fourth truck merged onto the road between the two decoupling platoon trucks, then tried to pass the lead platoon truck and almost pulled out in front of the truck passing the second platoon truck. This suggested that it may be better to keep platoons together near on-ramps. Aerial footage also showed a number of nonplatooning trucks following each other closely.

Other observations were that platoons were often passed by other vehicles (largely due to the requirement that platoons strictly obey speed limits, not flow with traffic), and that platoons could stay coupled only about 30% of the time (again due largely to restrictions placed on platoons by the exemptions). Trucks were connected using automated braking (Level 1) and DSRC. Lessons learned were documented and can be downloaded from the European Truck Platooning Challenge website (https://www.eutruckplatooning.com/default.aspx).

Alkim mentioned some potential benefits of truck platooning: increased throughput, fuel savings, and reduced emissions. Truck automation could potentially reduce labor costs in the long term if trucks can be driverless. Further work on vehicle connectivity and automation will be pursued in a large EU research initiative called Horizon 2020. There is interest in cross-brand platooning. All trucks in the European Truck Platooning Challenge consisted of trucks from the same manufacturer.

Shladover, Director of the PATH Program at the University of California at Berkeley, attended the European Truck Platooning Challenge and he provided his observations. He was impressed by how quickly the many different players were able to reach agreement and organize the challenge (from project conception to execution in 6 months). He asked whether a similar platooning demonstration could be conducted in the United States, and noted that it would require agreements between the diverse players, such as AASHTO, truck manufacturers, shippers, labor unions, and others. In Europe, for the Truck Platooning Challenge, it helped greatly that the vehicle manufacturers association and the road operators association each worked to get their members to reach agreement. Shladover also noted that the lack of sufficient standards is also a hurdle. Various standards organizations (ISO, SAE, and ETSI) are working on standards and sharing drafts to coordinate and harmonize standards, but a lot of technical work is needed to complete these.

Recent Studies from the Organization for Economic Cooperation and Development on Truck Automation Policy Options and Labor Market Effects

Voege of the OECD International Transport Forum discussed a project to identify important policy issues and to develop a policy roadmap for road vehicle automation. He identified many questions and open policy issues, which he grouped under the following major topics:

- Infrastructure needs:
  - How will merging work, especially with long platoons?
  - Will platoons need safe stopping areas large enough for the entire platoon?
  - How might driverless trucks be used? Will they travel between logistics centers?
- Weights and dimensions:
  - Should limits on weights and dimensions be considered together?
  - Larger, heavier trucks can reduce emissions (potentially even more than automation or platooning), but how to address public and environmental concerns about oversize, or extra heavy trucks? Should they be restricted to certain roads (e.g., as in Australia)?
  - Should platoons be restricted to certain roadways?
- Safety and security:
  - Public acceptance will be important. How reliable is the technology, and how safe does the public perceive it to be?
  - Safety of other vehicles, pedestrians, and cyclists must be considered as well.
- Capacity and modal shift:
  - Would increasing road capacity induce additional travel by passenger cars?
  - If truck platooning reduces trucking costs, will freight movement shift from rail to truck?
- Labor market effects:
  - At high automation levels, the driver workload will be greatly reduced, but this will be economical only if hours of service regulations allow longer driving–working hours.
  - What might high levels of automation look like? Perhaps remotely operated vehicles–platoons, but eventually driverless trucks?
  - With full automation, fewer drivers will be needed. Young drivers will need retraining, and older drivers may retire early (with pension implications).
- Urban freight:
  - Full automation will be very difficult to achieve in urban areas, however small, remote-control vehicles might be able to make deliveries.
  - How would such "road drones" and other new concepts be integrated with urban transportation and the urban environment?

The audience generally agreed that these are important questions, but many of them are impossible to address with current information, and the future of full automation of trucks and freight is very uncertain. However, as vehicle automation policies are developed, these questions should be considered.

Panel Discussion on Truck Cooperative Automation Issues and Opportunities

Richard Bishop led a panel discussion with a number of truck automation stakeholders. The panel included:

- Elizabeth Fretheim, Wal-Mart;
- Steven Boyd, Peloton Technology;
- Aravind Kailas, Volvo;
- Ed Hutchinson, Florida DOT; and
- Osman Altan, FHWA.

Panelists discussed why truck automation is interesting to their firms. Fretheim mentioned that Wal-Mart has a large fleet of Class 8 tractor trailers as well as many light- and medium-duty vehicles. Wal-Mart wants to make vehicle operation more efficient, but the top priority is service to the stores. Discussions with truck manufacturers on technologies are being held. There is interest in ADAS, but not driverless trucks. She mentioned the need to make systems easy to use for drivers, and standardization to make it easy to learn to use ADAS without having to relearn systems offered by different manufacturers. She stressed that technologies have to provide a return on investment.

Boyd described Peloton's efforts to bring partial automation and connectivity to market faster to improve safety and efficiency. Using V2V, V2C, and V2I connectivity, they are working with professional truck drivers to learn under what conditions and on what roads platooning works best.

Kailas briefly described Volvo's interest in truck automation. Driver comfort is important to Volvo. He said that the right automation, used the right way will have environmental benefits as well as benefits to drivers. Kailas mentioned that public awareness of seems to be increasing.

Altan mentioned the importance of freight operations overall and the large impact of trucking on the U.S. economy. He mentioned that labor and fuel costs are the largest components of truck operating costs. Truck automation can increase safety, which is the FHWA's number one priority. Automation can also increase capacity and throughput and decrease fuel use and emissions. He mentioned several projects FHWA has on truck platooning, with trucks using Level 1 automation (longitudinal control) and DSRC. Projects are evaluating the forming, maintaining, and decoupling of platoons, how truck platoons interact with other vehicles, driver preferences, and other human factors. Tests on a track will evaluate energy and emissions benefits. Altan said that FWHA would like to examine Level 2 automation and V2I technologies. Other topics Altan mentioned were managed–dedicated lanes for truck platoons, platooning with trucks from different fleets, and traffic harmonization.

Hutchinson mentioned that lots of goods are moved into Florida with their large tourist population and little manufacturing. Freight issues include the need for truck parking, and the shortage of drivers. Florida has a state law requiring a 300-ft separation between trucks on the road and prohibits TV monitors in trucks. However, the state legislature ordered a truck automation study, so Florida DOT wants to conduct a truck platooning pilot. There have been discussions between the Florida DOT infrastructure, operation, signage personnel, and state structural engineers about bridges and other infrastructure, and the Florida Motor Vehicles Department about potential licensing, permitting, and enforcement issues. He mentioned a potential evaluation could involve automation applied to trucks moving flowers between the Miami airport and a distribution center.

Regarding the recent interest in truck platooning in particular, Altan observed that with more technology being offered on new trucks, such as automatic transmissions, brake-by-wire, and electronic speed control, the incremental cost of technology to enable platooning is decreasing.

Fred Wehrli mentioned that truck platooning and automation in general are consistent with the Army's operational concept of human–machine teaming. In the long term, they would like to remove drivers from hazardous situations by using driverless vehicles.

Participants discussed the potential of conducting a platooning demonstration in the United States analogous to the EU Truck Platooning Challenge. The topic is being considered by several organizations. Florida is working with Georgia and Alabama on Bluetooth sensors along roadways. The I-95 Corridor Coalition recently held a workshop on CAVs. The I-10 partnership (Arizona, California, New Mexico, and Texas) has expressed interest in truck platooning and CVs.

On automation of low-speed truck operations, it was noted that it is hard to imagine driverless yard trucks—someone still needs to hitch and unhitch the trailer. However, technologies will be adopted where they make sense and this will depend on the application and may even vary geographically. Others possible applications mentioned by participants included automation of drayage trucks, or operation of platoons on streets, e.g., with traffic signal priority for platoons.

Participants discussed approaches states, MPOs, and cities can take with truck automation. It was noted that several Smart Cities Challenge proposals include truck automation elements.

A discussion on full automation and architecture aspects provided more focus on engineering issues. This portion of the breakout session began with a presentation by McKay, RAND, on recent work by the U.S. Army on automated trucks. The Army would like to automate trucks to reduce the number of drivers required for convoys, e.g., have a driver only in the lead vehicle, with driverless trucks following, and eventually have the whole convoy consist of driverless trucks. He mentioned that driverless trucks have been used under controlled environments such as a Rio Tinto mine in Australia that used 79 fully automated dump trucks. Additionally, John Deere has demonstrated fully automated vehicles for agriculture operations. McKay mentioned an Army demonstration at Ft. Hood with three fully automated (driverless) trucks driving in an urban setting. Another demonstration evaluated high-speed automated convoy operations in different weather conditions. McKay described needs and conditions that are unique to commercial or Army truck operations and those that are common to both:

- Commercial: improved fuel economy desired, can use existing infrastructure, conform to U.S. regulations.
- U.S. Army: reduce risk to personnel desired, limited or no infrastructure, rough terrain, austere/hostile conditions.
- Common: need for communication, HMI, cybersecurity, sustainment–maintenance.

The study is analyzing the risks and opportunities of vehicle automation for the Army as well as examining operational implications for automation, such as training, maintenance, etc. By considering a number of case studies of attacks on Army convoys, some important risks have been identified. Vehicle communications, including connectivity and vehicle positioning are fundamental requirements, and redundant/robust communications are needed to operate even under degraded conditions. The HMI should allow for optional driver operation, but there are design constraints on the useable space in the cab. New technologies have uncharacteristic failure modes, and users do not know how to deal with these failures. Cyber security threats are significant, and the "threat surface" is increased with sensors and communication links. Sustainment and maintenance considerations include mission availability and need to increase maintenance expertise. A practitioner's workshop was also held to learn from drivers who had been in the Army in Afghanistan and Iraq to understand how users interact with technology when performing tasks. They will issue a report in the fall of this year.

There was further discussion about commonalities and differences between technology for truck automation for commercial applications and military applications. As with commercial applications, the Army relies on commercial technology suppliers for sensors, processors, and actuators. V2I connectivity is less important for military applications than for commercial applications. Although military trucks have different requirements, there are stringent needs for verification and validation (V&V) of software and controls. V&V is very challenging for fully automated vehicles (both military and commercial). The Army has done some work on reference architecture standards, which can enable modular system design. Modules that can be separately tested and validated can reduce some V&V challenges as well as enable standardization of design elements. The Army is working with the other U.S. military services to develop joint requirement documents for vehicle automation. The Army is also interested in learning from commercial fleets about best practices for preventing adversaries from entering cargo spaces.

Contrasts between commercial truck automation and passenger vehicle automation were mentioned, namely that more advanced automation and connectivity technologies are being deployed on passenger vehicles that on trucks. Additionally, since there are many more passenger cars than commercial vehicles, connected cars with sensors are collecting much more data.

The supply chains for military, passenger and commercial vehicles are different, and business cases for automation will differ for these types of vehicles. The Army needs very high levels of automation much more than driver assistance, where the opposite is true (for the near term) for passenger and commercial vehicles. The Army tends to keep vehicles for 20+ years, so it is interested in maintainability, interoperability, and retrofitability more so than passenger car manufacturers. However, the military needs technology to be affordable, so it is important to be able to use commercial technologies rather than develop custom solutions.

MOVING FORWARD

- Participants discussed opportunities for collaboration between the private sector and the military, as well as between commercial trucks and passenger vehicles.
- Participants thought that examining HMI aspects of the evolving driver role with truck platooning is essential.
- Exploring near-term high-automation truck platooning deployments in limited geographic settings was additionally noted as important by participants.
- Participants suggested that continuing to focus on the interoperability of cooperative systems in the long-term was also important.

APPENDIX H: BREAKOUT SESSION 8

Traffic Signal Control with Connected and Automated Vehicles

Organizers

- Henry Liu, University of Michigan; and
- Larry Head, University of Arizona.

Reporters

- Henry Liu, University of Michigan;
- Larry Head, University of Arizona; and
- Katherine Turnbull, TTI.

SESSION FOCUS

This session focused on the impact on traffic signal control systems from CAVs. The possible impacts from a very low percentage of CAVs to a higher percentage of CAVs and eventually to fully autonomous transportation systems when traffic signal controls may no longer be needed were discussed by session participants. The PowerPoint presentations provided by some of the speakers are available on the AUVSI website at http://www.automatedvehiclessymposium.org/home.

SESSION SUMMARY

The breakout session included the following four presentations:

- **Multimodal Considerations in CAV Traffic Control**. Larry Head, University of Arizona. This presentation examined if traffic signals will be needed with a fully autonomous vehicle system and the transition during market adoption. The Multi Modal Intelligent Traffic Signal System (MMITSS) project was discussed. The project sponsors, basic concepts, and components were highlighted. The various elements of an integrated priority traffic signal control model were discussed, including the modal preferences and weight analysis, the coordination priority model, and the extended priority model with queue consideration.
- **Next Generation Traffic Signal Systems with CAVs**. Henry Liu, University of Michigan. This presentation addressed existing traffic signal systems, SMART signal system development, and future systems with CAVs. The Safety Pilot Model Deployment in Ann Arbor, Michigan, was described. The transition to the next generation of traffic control systems was explored, including a simulated signal-free intersection with self-driving vehicles.
- **Core Technologies for Cooperative Vehicle Adaptive Signal Control Systems** (CoVASS). Young-Jun Moon, Korea Transport Institute. This presentation described the CoVASS project funded by the Japanese Ministry of Land, Infrastructure and Transport (MoLIT). The MoLIT road vehicle policy, which includes reducing traffic fatalities, and the road

vehicle automation policy were summarized. The CoVASS control strategies and algorithms were outlined. Future activities were discussed.

- **Intersection Vehicle Infrastructure Cooperative Traffic Management and CAVs**. Wei-Bin Zhang, California PATH Program. This presentation explored the evolution of traffic control strategies, the California CV Testbed on El Camino Real in the San Francisco area, and traffic signal control to improve safety focused on red light running. Other topics covered included eco driving and traffic signals and intersection vehicle infrastructure cooperative CAVs.

MOVING FORWARD

Participants discussed research opportunities and issues regarding traffic control with CAVs. The following topics were identified for addition research: network level optimization, user characteristics, institutional issues, traffic flow theory with CAVs, application scenarios, signal control algorithms, establishing a baseline for research analysis, infrastructure adaption, examining transition periods to the next generation of traffic control, and the impact of shared mobility on traffic control. The following expanded descriptions of these topics were developed by the session organizers after the session.

Network-Level Optimization

Research to examine extending traffic control at isolated intersections to a network level with different traffic demands: Traffic control should interact with vehicle routing, including lane choice of each individual vehicle. New control algorithms will need to be developed to model multimodal and heterogeneous traffic (e.g., different sizes of vehicles). To perform network-level traffic control, it will be necessary to partition the network into different control zones using a network fundamental diagram. The objective of the partition is to limit the number of vehicles entering each control zone, which act as access control. Signal control can also group signals into different zones.

User Characteristics

Research to characterize the level of traffic control intelligence based on criteria such as communication technology, algorithm, and penetration of CAVs. Similar to the level of automation for vehicles defined by SAE, different levels of intelligence in traffic control can be defined. Individual vehicle characteristics (e.g., acceleration and deceleration profile, emissions), level of vehicle automation, and connectivity should also be considered in signal control system design. In addition, signal priority can be provided based on individual vehicles. Vehicles with a higher value of time could purchase priority and ridesharing vehicles could acquire higher priority.

Institutional Issues

Research on existing public and policy maker opinions on CAVs and autonomous technology: Identifying methods to promote technology adoption to public agencies, other organizations, and to the public would also be beneficial. These approaches could include education and outreach programs as well as incentives for technology adoption.

Traffic Flow Theory with Connected and Automated Vehicles

Research is needed to investigate the changes in traffic flow resulting from CAVs and the corresponding impacts on traffic control. The intersection capacity may increase with the increasing penetration of CAVs and the control strategies under different levels of penetration rates will probably be different. When the market penetration is very low, the existence of CAVs may even have negative impacts because regular drivers may be confused on how to react with CAVs.

Application Scenarios

Research on different communication technologies in addition to DSRC is needed. The performance and economics of other communication technologies such as cellular could be compared to DSRC. The type of communication may depend on particular applications. For example, some traffic signal applications may not need low latency, high-frequency DSRC communication, but safety critical applications do. Research defining a communication requirement specification for traffic control would be beneficial. Application scenarios, such as eco-driving and speed harmonization, can be integrated together.

Signal Control Algorithms

Research is needed on designing robust signal control algorithms that are able to tolerate faults, including system failures. The control system should be adaptive to difficult conditions such as server weather. Since the trajectories of CAVs can be controlled, more measures can be used to determine the performance of the algorithms in addition to vehicle delay. Vehicle routing information is very important to real-time adaptive signal control and path-based optimization.

Establish a Baseline for Research Analysis

A common baseline for research analysis is needed. The baseline could be used to compare the performance of different CAV signal control algorithms. The baseline should include estimation of gains, improvements, and expectations at different levels of control. Factors to include are the market penetration rate, the level of intelligence in the control system, the constraints on the system, and other elements. The complexity or degrees of freedom should be carefully designed in developing a baseline. The trade-off between the number of parameters and use should be considered.

Infrastructure Adaption

Vehicle lanes may be managed to adapt CAVs on the roadway. Some lanes could be assigned as dedicated CAV lanes when the market penetration rate increases. Lanes can be dynamically assigned to different types of vehicles at different times if vehicles can be packed into groups by type. Research developing dynamic intersection maps to reflect the changes in lane assignment, stop-bar location, and other factors would be beneficial.

Transition Period

Data fusion is necessary when applying CAV-based applications when the penetration rate is low. Integrating traditional high-resolution signal operation data with CAV trajectory data provides the opportunity to improve the performances of current signal operations. When the penetration rate increases, trajectory data from CAVs should be enough to operate traffic signals.

Impact of Shared Mobility on Traffic Control

Most of the traffic signal control strategies today consider that the traffic demand is a given. In the future, transportation network service providers (e.g., large shared fleets owned by private companies) will generate most traffic demand instead of individual travelers. These companies try to optimize their own operations, which also interact with the traffic control system. Investigating this bi-level problem of the relationship between transportation network service providers and the traffic control system would be beneficial.

APPENDIX I: BREAKOUT SESSION 9

Methods for Assessing Market Acceptance, Adoption, and Usage of Automated Vehicles

Organizers

- Johanna Zmud, TTI;
- Arum Kuppam, Cambridge Systematics;
- Aruna Sivakumar, Imperial College London
- Barbara Lenz, German Aerospace Center, DLR;
- Guy Rousseau, ARC;
- Hyeon-Shic Shin, Morgan State University;
- Jana Janarthanan, Washington State DOT;
- Junyi Zhang, Hiroshima University;
- Kouros Mohammadian, University of Illinois at Chicago;
- Kristin Kolodge, JD Power and Associates;
- Maren Outwater, Resource Systems Group;
- Naohisa Hashimoto, National Institute of Advanced Industrial Science and Technology;
- Paul Leiby, Oak Ridge National Laboratory;
- Scott Le Vine, Imperial College London; and
- Siva Srinivasan, University of Florida.

Reporter

- Johanna Zmud, TTI

SESSION FOCUS

This session focused on how to collect reliable and accurate information on market acceptance, adoption, and usage when people generally have little knowledge of, and no experience with, AVs. The session was conceived as an interdisciplinary dialogue on how best to study the ways in which the market for AVs will develop. Because AVs are not yet present in the traffic streams, with the exception of a few test vehicles, it is difficult to reliably predict future consumer demand. Any purported outcomes are just theoretical at this point. The choice of methods and the way in which they are implemented are largely determined by the research questions, but are also influenced by practical considerations such as ethics, budget, and time. In this workshop, transportation planners, researchers, and policy analysts exchanged their research experiences to date, identified priority (near-term) research questions, and explored best practice for the future. To simplify the challenge, the session emphasized vehicle ownership and usage uncertainties.

SESSION SUMMARY

Opening Remarks

Johanna Zmud, *TTI*

- The session objectives are to determine and prioritize the information required to understand vehicle usage and ownership uncertainties and to outline suggested research approaches for capturing reliable information.
- The session salience was based on the facts that AVs are new technology with which consumers have no experience. This situation creates new challenges in examining usage and ownership questions. People cannot their responses to questions based on past or current experience. So how do we proceed?

The agenda was comprised of four speaker presentations, followed by small facilitated breakout groups and reports on those discussions.

Speakers	Summary of Presentation and Question and Answer Session
Mike Quinn, Boston Consulting Group	**Revolution in the Driver's Seat** • Discrete choice modeling of 2035 global market penetration of AVs based on a survey and expert interviews, together with scenarios to account for influencers on penetration. • It was predicted that global AV shares will reach 9.6% with full AV functions, 15.6% for partial AV functions. • Many people's willingness to pay was US$5.00 for AV functions.
Stefan Trommer, DLR	**Exploring User Expectations of Autonomous Driving** • Focus groups to develop scenarios to derive inputs to vehicle fleet and travel demand models for U.S., Germany, and China. • Focus groups enabled an assessment of benefits and potential use of AVs for different user groups within different countries. • Qualitative approach was key for discussion of potential changes in mobility behavior: – Knowledge about AVs within the focus groups was very limited. – Discussion allowed participants to generate a mutual understanding. – Understanding is prerequisite to evaluate potential utilization of AVs.

Speakers	Summary of Presentation and Question and Answer Session
Dan Weinstein, Resource Systems Group (RSG)	**Forecasting Consumer Adoption of Automated Vehicles and the Impact on Total Vehicle Sales** • Joint revealed–stated preference modeling approach to test sensitivity to owned AV and pay-per-use (PPU) attributions and to model adoption and sales impact. • Revealed preference data of interest includes vehicles in household, and bring those specifically into choice exercises; VMT per existing vehicle; and existing usage of PPU ride hailing/sharing. • Stated preference to gather preferences on AV purchase features such as cost increment versus traditional, time on market, and accident rate. • PPU information on such as cost per mile, average wait time, maximum number of passengers, vehicle condition, and additional travel time incurred (pooled PPU).
Maren Outwater, RSG	**Summary of Results of Pre-Symposium Survey on Priority Research Questions Among Break-Out Session Leaders** • The most important vehicle usage variables to measure: – Acceptance, – Demand, – Cost, – Privacy, and – Quality of life. • The most important vehicle ownership variables to measure: – Adoption, – Cost, and – Automated features of interest.

The 50 participants were divided into four small groups; two each focused on vehicle ownership and vehicle usage issues.

- Key points from vehicle ownership discussion:
  - Challenge in researching is that there are multiple levels of uncertainty—technology, policy, and consumer preferences.
  - What can we say about now? How long will people keep the car they have at the moment, decrease of cost over time, and number of potentially additional users (e.g., handicapped, elderly).
  - What are near-term questions? What types of people prefer to own AVs? What are the limits of shared mobility? Why do people want to own a car? And how and how quickly will state agencies contribute to the implementation of AVs?
  - Gaining of experience before AV purchasing will be key. How to accommodate this in survey methods?
  - Pilot studies are needed. Perhaps using behavior analogs such as providing a "normal" car and chauffeur to simulate a similar mode as Level 5 automation.
- Key points from vehicle usage discussion:
  - We cannot predict adoption behavior before AV technology is available (and understood).

– Existing household travel and time-use surveys are not sufficient to anticipate new trips and new purposes for which AVs may be used.
– Majority of research still highly uncertain especially value of time and willingness to have passengers.
– Lots of noise in current research. Recommend to study continuously over time to determine trends and potential for dynamics.
– Given that we will not forecast it correctly, it does not mean doing nothing. We need to inform the industry as best we can.

CLOSING DISCUSSION COMMENTS

Key Takeaways and Lessons Learned

- Vehicle usage:
  – Key unknowns: public acceptance, incentives, and disincentives to usage (private versus shared), value of time, across purposes, across modes.
  – Study continuously over time to determine trends and potential for dynamics.
  – Bring creativity—new methods to anticipate and capture behavior change (new trips, purposes, influences on lifestyle, and time use).
  – Understand need to "educate" respondents prior to actual data collection.
  – Find behavior analogs (e.g., vouchers for chauffeurs).
  – Leverage any pilot tests to carefully study user behavior (with and in-vehicle).
- Vehicle ownership:
  – Need to understand usage to understand ownership.
  – Key unknowns: willingness to pay, ownership persistence, size, and impact of "new" owners.

MOVING FORWARD

- Continue the dialogue.
- Next year: plan exercise that presents a hypothetical research sponsor with a particular decision to make—what are innovative methods–approaches to get required information.
- Educate decision makers, funders; creativity costs money.
- Collaborate on new research: mixed method, cross-disciplinary, and cross-national.
- Track, compile, and promote examples of behavior analogs, creative methods, etc.

APPENDIX J: BREAKOUT SESSION 10

Ethical and Social Implications of Automated Vehicles

Organizer

- Patrick Lin, California Polytechnic State University

Reporters

- Suzie Lee, VTTI; and
- Katherine Turnbull, TTI.

SESSION FOCUS

This session examined potential ethical and social issues associated with AVs. The session focused on the importance of ethical considerations, how developers may think about values in programming decisions and managing risk, ethical issues in licensing and testing, and consumer perceptions. Some of the session panelists provided prepared comments, which are available on the AUVSI website http://www.automatedvehiclessymposium.org/home.

MEETING SUMMARY

This session included four panels. The panelists provided short remarks, followed by questions and discussions with all the participants. The session was the first time the topic of the ethical and social implications of AVs was included in the AVS breakout groups. The comments by panelists providing prepared summaries are highlighted.

Panel 1: Why Ethics?

The panelists included Emily Frascaroli, Ford Motor Company; Stephen Wu, Silicon Valley Law Group; and Tom Powers, University of Delaware. Patrick Lin, California Polytechnic State University, moderated this session.

- Wu and Powers provided written comments. Wu discussed potential legal issues with AVs related to compliance, liability, and information governance. He described three possible traditional defenses to consider if a manufacturer can avoid liability in what might appear to be an ethical design choice—necessity, defense of others, and the sudden emergency doctrine. Powers outlined a potential research program in machine ethics based on the concept of coded ethics. This approach would identify the ethical principles that are explicit in codes of ethics and formalize them into principles that can be written into software that controls AVs and other related technologies.

Panel 2: Values and Weights

This panel included Noah Goodall, Virginia Transportation Research Council; Stephen Erlien, Peloton Technology; and Sarah Thornton, Stanford University. Ryan Jenkins, California Polytechnic State University, served as the session moderator.

- Erlien provided written comments. He discussed the concept of uncertainty and how it links to quantifying risk. He described uncertainty related to AV decision making and the factors that influence uncertainty.

Panel 3: Licensing and Testing

Wendy Ju, Stanford University; Suzie Lee, VTTI; and Shad Laws, Renault Innovation Silicon Valley were the panelists. Keith Abney, California Polytechnic State University, moderated the session.

- Ju and Lee provided written comments. Ju discussed the need for the on-road testing of autonomous vehicles, analyzing the results, and sharing collected data. As an interaction design researcher, she noted the importance of considering how people interact with other people, with things, with services, with culture, with information, and with the world. Lee discussed two human-subjects tipping points for consideration in licensing and testing AVs. The first point was the experimental approach of introducing people to testing AVs with the "come out and try it—it's great" approach. The second point was introducing vehicles or subsystems into the market without human subject research. She also reviewed key elements of human subject research with AVs.

Panel 4: Consumer Perceptions

Sarah Hunter, X; Joe Barkai, IDC; and Jason Millar, University of Ottawa. Patrick Lin, California Polytechnic State University, moderated the session.

- Barkai and Miller provided written comments. Barkai discussed factors that influence technology adoption, including AVs. He suggested that the notion of trust in technology should be added to the discussion of public perceptions of AVs. Miller discussed the role of public engagement in the design and engineering of AVs. He described the use of surveys to obtain public input on ethical questions related to AVs.

OPEN DISCUSSION

- Participants discussed that there are no easy answers to ethical dilemmas, and that no answer is black and white. Ongoing discussions were suggested to examine key issues.
- Participants suggested that ethical decisions are not being hard codded into AV systems. It was further suggested that based on the optimization scheme (safety principles, efficiency principles, energy principles, or other principles), the actions taken by the vehicle may appear to be making a decision based on ethical considerations.

- Participants discussed the potential to leverage advances made in bioethics over the past 50 years to help address the potential issues with AVs.
- Participants suggested that OEMs and developers should employ ethicists to ensure that these discussions are occurring during the design phase, going above and beyond the minimum legal standard of care, which would help alleviate possible product liability concerns.

MOVING FORWARD

- Participants discussed that standards are needed if data is to be shared. The standards would address the type and amount of data, the format, and other elements.
- Participants suggested that on-road testing was needed after computer modeling and simulations. It was also noted that human subject protection considerations will need to be addressed with tests and pilots.
- Participants supported continuing the session at future symposiums. Possible topics for future meetings included lessons learned from on-road testing, lessons learned from any crashes and incidents, and new ethical considerations based on new technological advances.
- Participants suggested that ongoing critical thinking was needed on the topic.

APPENDIX K: BREAKOUT SESSION 11

Early Implementation Alternatives
An Interactive Scenario-Planning Session

Organizers

- Caroline Rodier, University of California, Davis;
- Marco Anderson, Southern California Association of Governments;
- Sahar Shirazi, California Governor's Office of Planning and Research;
- AnnaLisa Meyboom, University of British Columbia;
- Chris Ganson, California Governor's Office of Planning and Research;
- Francesco Ciari, Swiss Federal Institute of Technology, Zurich; and
- Regina Clewlow, moovel Group.

Reporters

- Caroline Rodier, University of California, Davis; and
- Marco Anderson, Southern California Association of Governments.

SESSION FOCUS

AV technology holds the potential to revolutionize mobility as we know it today. How and when this revolution might occur, is subject to uncertainties related to technological feasibility, consumer perceptions, and institutional barriers. Breakout Session 11: Early Implementation Alternatives: An Interactive Scenario Planning Session drew on qualitative scenario planning methods to explore the implementation of near-term applications of AV technology (NHTSA 1-4) in specific geographic contexts and time horizons. The moderators provided plausible future contexts for implementation of AV systems, which included complete streets, first- and last-mile transit access, and managed lanes. Participants also received questions and prompts to help them as they explored each scenario. Participants broke out into groups of 10 to 12 people who had expressed preference for similar scenario problems. Participants were also asked to prepare a 15-min presentation to share with the larger group.

SESSION SUMMARY

Complete Streets Scenario

The year is 2040 and the city of Los Angeles has carefully implemented the principles of complete streets in their downtown since the beginning of the 21st Century. A bikesharing program was introduced in 2015 and is still very popular. Today, drivers, bicyclists, and pedestrians have equal access to the street-level infrastructure. Market penetration of personal autonomous vehicles has accelerated rapidly over the past 5 years, which has exponentially increased the capacity and vehicle throughput on the highways leading to downtown Los Angeles. As a result, the streets are now approaching near gridlock, which is compromising culture and businesses downtown. The city of Los Angeles has recently decided to limit vehicle access to a fleet of small, electric, autonomous shared-ride (or shared-taxi) vehicles. No personal AVs or other vehicle services will be allowed in the downtown urban core. AVs don't require street-level parking and will use very little of the current parking capacity in the city. Demand is high for new downtown space, and there is huge interest in converting parking structures to more profitable uses.

The complete streets group began by identifying the objectives of the project. The group agreed that the project should provide a higher level of transportation service to more people without increasing roadway capacity. They asserted that this would benefit businesses and customers without increasing roadway infrastructure costs.

The group proposed a shared AV service that would be run by one or more fleet operators (public or private entities). The service could range from demand-responsive shared-ride services with small vehicles to continually circulating shuttles. To encourage public acceptance and support for the plan, the group suggested a transition period with trials on certain cordoned or selected street closures.

Concern was also expressed about access for the disabled. It was decided that the disabled would be given a special pass to drive an AV into the zone without cost. The group acknowledged that this would require some kind of enforcement mechanism.

There was general agreement among group members that transition zones at different entrances of the project areas would not be practical, efficient, or acceptable to the public (e.g., imagine airport parking). Instead, they asserted that such a scenario would only be possible with very high-quality public transit service to key entrance zones of the project area. In essence, this scenario became a specialized first-and last-mile problem. High-quality first- and last-mile service would also be needed at the home end of this transit service, which could be also be met with a shared ride AV service.

The group anticipated significant loss of city revenue from metered on-street and off-street parking. Incremental pricing (i.e., increasing property tax when property gains more value) was raised as a possible solution. However, it was pointed out that California law prohibits this type of taxation. Another idea was to charge a fee for use of the shared ride AV service. Others preferred the imposition of a city or county sales tax. They argued that the service would be of significant benefit to local businesses and thus a half-cent sales tax would be justified.

The interaction of AVs with pedestrian and bicyclist traffic was also identified as a significant problem. One idea was to give traffic priority to pedestrians and bicyclists. However, others raised the concern that this could result in long AV queues. The group was also concerned that the proposed service would require curb space for frequent drop-off and pick-up travel, which would require the use of old on-street parking spaces and would conflict with the use of typical outside bike lanes. As a result, they suggested relocating bike lanes on the inside of lanes. The group noted that left the issue of the movement of bicyclist to the outside lane at certain turn intersections unresolved.

First- and Last-Mile Access Scenario

The year is 2025 and the South Coast region has successfully implemented an efficient transit rail network that provides high-quality linkages among the cities and major employment and population centers. Regional transportation planning was successful in increasing densities within a quarter-mile of transit stations, but beyond this area most of the housing development takes the form of typically low-density suburbs. Parking at rail stations in suburban cities now fills up at 6 a.m. and after that, the rail service is only lightly used until the afternoon–evening peak hours. This usage is limiting fare revenues, which are sorely needed. An active bikesharing program has extended the perimeter for some commuters who are about 5 mi away from the station and are comfortable biking to the transit station. Transit agencies have tried to implement first- and last-mile service to the rail stations, but in order to keep fares at a reasonable price point, bus travel times to the stations can be as long as 30 to 40 min. Some promising demonstration projects in Europe and Asia show the potential for low-speed AVs to provide timely and affordable first- and last-mile service. The cities and counties are now joining together to a similar service in the Los Angeles region.

The first- and last-mile group developed a service concept that involved low-speed (35-mph) AVs that would share dedicated lanes with bicycles. Bicycles would have priority over AVs. This would form a hub-and-spoke system around suburban high-speed transit. The system would include an automated microtransit shuttle type service (e.g., Bridj) for larger groups (10 to 12), an automated taxi service (e.g., Uber or Lyft) for single or smaller groups (one-to-six passengers), and an ebike shared-use service. During the off-peak periods, microtransit could perform relocations services for the ebike and provide delivery services for local businesses. Existing station parking would be redeveloped to accommodate increased residential and employment densification around the transit station. Access to the train station would be redesigned to include necessary curb space for drop-off and pick-up travel as well as the ebike sharing system.

The group also identified a number of possible method for funding this project:

- Potential profits from redevelopment of parking lot space;
- Charge for remaining parking;
- Governments funds such as bonds;
- Co-op ownership (e.g., investor puts some money into the system and owns part of it);
- Revenue from off-peak delivery services;
- Carbon tax revenues; and
- Countywide sales taxes.

Autonomous Road Trains in Managed Lanes Scenario

The year is 2020 and transportation agencies throughout southern California have planned and worked toward a network of managed lanes. These lanes were originally designed to include only HOVs and incentivize carpooling and vanpooling by providing faster travel times compared to general-purpose lanes on the same route. However, rapid urbanization and population growth have resulted in widespread congestion in the HOV-only lanes. AVs are technically viable, but market penetration has been slow. Agencies are now working together to consider how to expand the capacity and reduce congestion in these managed lanes by introducing AV "road trains." It appears that this may be a good fit because managed lanes are separated from the general-purpose lanes and can accommodate electronic tolling, which can be used to fund connected AV infrastructure and technology to maximize capacity and provide fuel-reduction benefits.

The autonomous road trains group started out by establishing a set of goals and objectives for the project. These included the following:

- Increase the adoption rate of automated vehicles;
- Improve the environment (especially, GHG emissions);
- Ensure equity of access for all travelers regardless of their income;
- Improve economic efficiency (i.e., increase the overall throughput of the freeway); and
- Attain safety rates that are comparable with airline travel.

This group was very worried about the public acceptance and political feasibility of this concept. As a result, they imagined that this project would be designed for a federal- or state-funded demonstration project (e.g., Urban Partnership Agreements and Smart Cities). In addition, it was decided that it should be implemented in areas with high levels of congestion and where there were physical barriers that prohibited freeway expansion (e.g., I-15 in San Diego). The focus would be on the long-distance commute market. The lane would allow for shared-ride AVs, automated buses, and personally owned AVs. Those who did not travel by bus or share rides would be subject to a toll. The revenue from this toll would be used to subsidize shared riders for low-income travelers and the disabled. The system would require platoon-ready AVs at Level 3 and above with V2V communication, adaptive–smart cruise control, and automated breaking.

The group discussed many implementation issues related to responsibility for system operation, how to establish a platoon leader, data-sharing agreements, credentials, and security management. However, in the end the group felt that the interaction of automated and nonautomated vehicles was the most significant barrier to implementation and could render the project, given its goals, infeasible. Entering and existing the automated lanes from the nonautomated lanes may make it extremely difficult to maintain spacing and speeds goals. Because long platoons with similar destinations may be difficult to form and would fill up the capacity of managed lanes, the group contemplated reducing the length of trains to two or three vehicles. However, this may make entrance and exits more vulnerable to slower speeds and reduced capacity.

In summary, the cost of implementing managed lanes may be high and careful research would be beneficial in understanding market demand and transition effects on the capacity of managed lanes. In general, if the goal is faster adoption of AVs, other approaches may be more effective, for example, subsides for using shared-ride AVs services and cash-out programs for nonautomated vehicles.

CONCLUSION

Participants engaged in lively debates about the challenging issues surrounding selective implementation of AV systems in the short time available during this session. A number of critical challenges related to the interaction of AVs, nonautomated, intercity transit, and bike and pedestrian traffic were identified. Careful multidisciplinary research would be beneficial in exploring how these interactions can be managed. Small-scale tests of technical and functional feasibly may be reasonable short-term project early implementation projects.

APPENDIX L: BREAKOUT SESSION 12

Automated Vehicle–Ready Cities or City-Ready Automated Vehicles

Organizers

- Stephen Buckley, City of Toronto;
- Philippe Crist, International Transport Forum;
- Siegfried Rupprecht, Rupprecht Consult; and
- Jane Lappin, Toyota Research Institute.

Reporters

- Siegfried Rupprecht, Rupprecht Consult; and
- Katherine Turnbull, TTI.

SESSION FOCUS

This breakout session addressed key aspects of road automation from an urban policy perspective. Speakers and participants examined how cities can create an environment to ensure that CAVs deliver the anticipated safety and mobility benefits. The PowerPoint presentations provided by some of the speakers are available on the AUVSI website at http://www.automated vehiclessymposium.org/home.

Opening Remarks

Jane Lappin, *Toyota Research Institute*

- Lappin noted that the urban perspective is prominently on the AVS agenda. This breakout session is a first attempt to create a dialogue between urban stakeholders that needs to be continued.

Liam Breslin, *European Commission, Directorate General Research and Innovation*

- Breslin reported that European research programmes such as Horizon 2020 are addressing a wide range of technological, human factor, and policy impacts of road automation. Projects such as CityMobil2 have successfully evaluated the impacts of automation on public transport in cities.

SESSION SUMMARY

What Cities Need: Urban Mobility and Road Automation	
Speakers	**Summary of Presentation**
Philippe Crist, International Transport Forum	• Crist discussed the different approaches to vehicle automation strategies, including having some technology everywhere, having all technologies somewhere, and having all technologies everywhere. He described the rapid development of all types of technologies. • Crist described an analysis of the impact of a shared, fully AV fleet in Lisbon, Spain. The analysis used an agent-based model simulating the behavior of all groups in the mobility system. In a 24-h scenario, only 3% of the current number of cars was required to provide the same number of trips. In addition, 20% of the on-street parking spaces and 80% of the off-street parking facilities were not needed. Vehicle kilometers traveled were reduced, vehicle emissions declined, and access to jobs increased. • Possible policy insights include the need for a strong but flexible regulatory framework, as well as the need for public transport, taxi operations, and governance to adapt. The potential benefits are linked to system and market design, as well as managing available space from the reduction in the need for parking.
Natalia de Estevan-Ubeda, Transport for London	• de Estevan-Ubeda described the role of Transport for London (TfL), challenges the city is facing, the new mobility landscape, and the deployment of CAVs. She reviewed the diverse transport system in the city, which includes roads, buses, taxis, the London Underground and Overground, dial-a-ride systems, river ferries, bicyclists, pedestrians, and other modes. The TfL implements the Mayor's Transport Strategy in the context of integrated strategic planning for land use, transport, and economic development. • de Estevan-Ubeda described the challenges facing London, including population and employment growth, which are placing additional demands on the transport system. She also discussed some of the market disruption trends, including technology advancements, changing lifestyles, shifting customer expectations, and the blurring of public and private transportation boundaries. She also highlighted some of the potential unintended consequences and dis-benefits from CAVs. • de Estevan-Ubeda discussed the potential scenarios for CAV deployment, possible proactive and reactive agendas cities could follow, and preparing for CAVs in uncertain times. She noted that TfL is exploring an integrated strategy considering a wide range of factors.

What Cities Need: Urban Mobility and Road Automation (*continued*)	
Speakers	**Summary of Presentation**
Seleta Reynolds, Los Angeles Department of Transportation (LA DOT), California	• Reynolds described a transportation technology strategy for Los Angeles. The vision of LA DOT focuses on providing a transportation system that gives people choices to support a high quality of life and strong healthy communities, as well as continued prosperity and resilience for the region. • Vision Zero, the Great Streets Program, and mobility management are all part of this vision and promoting equity. • Streets, freeways, and parking are major land uses. • The future vision is that LA DOT will support 21st-century technology for increased mobility, safe access, and sustainability and create truly great streets. • A platform for mobility innovation is being developed. Key elements are building a solid data foundation, leveraging technology and design, creating partnerships, supporting continuous improvement through feedback, and preparing for automation. • The platform also focuses on data as a service, mobility as a service, and infrastructure as a service. Layered pilot schemes in all these areas are being developed and implemented.
Carla Bailo, Columbus, Ohio	• Bailo described the selection of Columbus as the winner of the Smart City Challenge. • The proposal included demonstration projects of mobility measures in different areas of the city. • Reducing congestion and improving safety/mobility on roadways in a commercial area, which includes shopping, businesses, and housing is one project. • An underserved residential area with numerous social challenges will be served by a BRT line with connections to first-mile/last-mile automated shuttles. • The downtown area will be served by BRT to connect citizens to jobs. • Truck platooning on arterials, combined with real-time data, will allow goods to move easily to their delivery points from the Rickenbacker facility. • The overall concept is to use technology and data to improve people's lives, which is what makes a "Smart City."
Mark Dowd and Brian Cronin, U.S. DOT	• Dowd and Cronin provided additional background on the U.S. DOT Smart City Challenge. A total of 78 cities applied to the Smart City Challenge. Seven cities were selected as finalists, with Columbus selected as the winner. • Columbus' success factor was that technology solutions were proposed for their key urban problems. • A transparent national process will be initiated to facilitate the exchange of information between cities and sharing of lessons learned. • The application process created interest and momentum among all the stakeholders involved.

How Can Cities Plan for Automated Vehicles? Scenarios of Automated Vehicle Introduction	
Speakers	**Summary of Presentation**
Emily Castor, Lyft	• Castor described Lyft, which is a transportation network company (TNC) providing on-demand ridesharing through smartphone apps. Lyft operates in over 200 cities in the country, with an expanding focus from leisure travel into mainstream mobility. • Automation will change the cost structures of TNC services. Integrating automation into future business models, as well integrating with mainstream public transport services to provide first- and last-mile connections represent opportunities and challenges. • Automated ridesharing services have the potential to reduce and potentially end car ownership. • The partnership with General Motors focuses on shared, electric, and AVs. • Automation will be probably be available only in certain areas or routes initially, but will be part of the wider marketplace of mobility options. A modest extra cost for automated services is expected.
Chris Boroni-Byrd, Qualcomm	• Boroni-Byrd discussed two possible scenarios for the deployment of AVs in cities. In one scenario, private cars are banned from the city center and lower-speed mobility modes are promoted. Creating trust in driverless mobility is important to support deployment. • In the second scenario, automation is introduced first in taxis and buses. The role of taxi drivers and bus operators transforms into a monitoring task. • Slow integration could begin now, as numerous benefits could be realized. Having fully automated vehicles is the endgame.
Steve Buckley, City of Toronto, Canada	• Buckley discussed some of the issues facing cities with the deployment of AVs, scenario planning AVs, and the approach being taken in Toronto, Ontario, Canada, to plan for AVs. He suggested it was important to elevate the discussion of how cities can and should shape the development of AVs, which may impact the design, development, and operation of cities. • Buckley noted the potential infrastructure, planning, regulatory, and economic issues cities may need to address with AVs. He described three possible scenarios with private AVs leading, shared AVs leading, and a mix of private and shared AVs. He suggested that a fundamental challenge was the unknowns associated with the speed of technological advancements, economics, and public acceptance. These unknowns make planning for the future difficult. He noted that cities could take different approaches from actively discouraging AVs, to a passive role, to actively encouraging pilots and deployment. • Buckley reviewed the activities and planning process underway in Toronto to consider AVs. An internal working group in Toronto includes representative from transportation, economic development, city planning, and licensing and standards. Other impacted groups are the police, parking, revenue, budget, information technology, and the privacy commission. He reviewed the draft vision statement, which is "Toronto needs to harness the potential of AVs to help us create the city that we want."

How Can Cities Plan for Automated Vehicles? Scenarios of Automated Vehicle Introduction (*continued*)	
Speakers	**Summary of Presentation**
Siegfried Rupprecht, Rupprecht Consult	• Rupprecht suggested that there may be quite a long period of coexistence between conventional vehicles and AVs and that transition challenges to an automated road transport system will be highest in cities. • City administrations and urban stakeholders are not sufficiently part of the current discussion about automation. Managing the urban automation transition well will be a key success factor for effective AV deployment overall, while avoiding new safety issues. • Automated urban transport of the future is most effective if it is a part of a quality collective transport system and an active demand management policy; lower personal car ownership will be a major success factor of automation. • Transport planning needs to deal with many uncertainties but should consider identifying corridors with gradually increasing automation support and pilot areas for shared automation. • Transport modeling is not yet "AV ready." • Building up "AV enabling" infrastructure is needed, including considering the needs of CAVs, VRUs, public transport, pick-up/parking areas, freight infrastructure, maintenance, and long-term development impacts. • A key future question will be how to redistribute space and to reassign urban functions? • New service models and government structures are required to agree jointly on an urban automation agenda. Policymakers should actively help develop an automation vision for their cities.
Group Discussion	• What can cities do to achieve their goals more effectively by creating a supporting AV framework? Participants discussed the following scenarios: ownership leads, on-demand leads, and ownership and on-demand split outcome. • Participants discussed that automation is getting on the urban policy agenda. They questioned if city policies are on the automation agenda of industry, academia, and government. • Participants discussed that automation is also a tool to implement housing, land use, and health policies, as well as addressing other critical needs in cities. Participants further discussed which transport problems automation can help to solve more effectively than other technologies or policies, such as clean fuel fleets and mobility as service. • The possible turning point where AVs are generating an impact on urban form was explored by participants. • Participants discussed potential challenges with automation in city environments. It was noted that avoiding unintended impacts, while dealing with uncertainties was one challenge. The need for planning tools and assumptions represented another challenge, as did the need to make long-term infrastructure decisions now while keeping pace with many technology advancements. • Participants discussed that the coexistence of AVs and conventional vehicles will likely be a long, and potentially messy, period. • Participants suggested that in order to benefit from automation, cities must proactively promote behavior change and create an enabling policy framework.

MOVING FORWARD

Rupprecht developed a potential urban automation agenda based on the discussion in the breakout session to begin an internal dialogue on actions from an urban mobility policy. Elements of this agenda included the following.

- Develop a vision of how automation can help transform mobility in cities, and how to promote behavior changes through shared-mobility services, quality transit development, and demand management policies.
- Support local shared vehicle pilots, including identifying impacts and upscaling when successful should be included in the agenda.
- Consider AV-enabled corridors using supporting policies and technologies and discuss how these corridors could develop over time and interact with other modes.
- Develop scenarios for reusing parking spaces and how transit infrastructure might evolve.
- Consider obsolescence, resilience, and new liability issues related to AV-ready infrastructure.
- Give special attention, including considering new service models and regulations, to the automation of freight and transit.
- Address public perceptions and involve stakeholders at all levels in developing AV-ready cities.
- Conduct research to develop a city-focused automation work plan with partners and stakeholders.
- Continuing to engage in joint learning and in dialogue with industry, academia, and government would be beneficial to all groups.

APPENDIX M: BREAKOUT SESSION 13

Design and Operational Challenges and Opportunities for Deploying Automated Vehicles on Freeways and Managed Lanes

Organizers

- Christopher Poe, TTI;
- Chuck Fuhs, TRB, Managed Lanes Committee;
- Casey Emoto, Santa Clara Valley Transportation Authority;
- Steve Kuciemba, WSP–Parsons Brinkerhoff;
- James Colyar, FHWA;
- Greg Krueger, HNTB Corporation; and
- Bob Ferlis, FHWA.

Reporter

- David Florence, TTI

SESSION FOCUS

This session focused on sharing recent research and studies examining the potential AVs use of managed lanes. The session helped to inventory what is known about how CVs and AVs could perform on managed lanes and to identify needs from an infrastructure and operational standpoint. To provide context and to help identify what is known about CV–AVs on managed lanes, the speakers shared emerging documents and case studies related to CV–AVs on managed lanes. The session also focused on identifying gaps in knowledge for AV deployment in controlled access facilities and to identify future research to be incorporated into collaborative statements of the sponsoring TRB committees.

SESSION SUMMARY

The session included seven speakers, followed by a facilitated group breakout discussion. PowerPoint presentations from some of the speakers are available on the AUVSI website at http://www.automatedvehiclessymposium.org/home.

Speakers	Summary of Presentation and Question and Answer Session
Ray Derr, TRB	**Overview of NCHRP 20-102: Impacts of Connected Vehicles and Automated Vehicles on State and Local Agencies** • NCHRP Project 20-102 was approved in December 2014 and authorized $3.5 million to begin work on the highest priority research projects. • One of the nine tasks in this project is to develop guidance and identify conditions for dedicated CV–AV lanes. • The project aims to use case study sites to develop and apply an analytical approach, identify laws and regulations, and develop guidance.
Joe Rouse, Caltrans	**Lessons Learned from Managed Lanes Applicable to CV–AV Dedicated Lanes** • Rouse identified access control and detection systems as managed lane elements that may be useful for CV–AVs. – Access control can make for easier deployment. – Detection system gantries could host readers and equipment to enable CV–AVs. • Access control would require detailed mapping and there must be identification of CV–AVs for ridership eligibility. • There is a potential issue with degradation of the performance with the addition CV–AVs. • CV–AVs may be able to help with occupancy verification and electronic toll collection.
Robert Ferlis, FHWA	**U.S. Department of Transportation Project Highlights** • CV–AV used as a tool for improving safety, increasing mobility, and reducing energy use. • FHWA has been working on timeline, which involves proof of concept, applied research, policy and deployment preparation and support, and initial deployment of CV–AV applications. • These applications include CACC, lane change–merge assist, speed harmonization, environmental optimization, and truck platooning.
Steve Kuciemba, WSP–Parsons Brinkerhoff	**Safety Pilot Model Deployment–Ann Arbor Connected Vehicle Test Environment** • Federally funded transition to operational development with deployment 1,500 CVs per year, with a goal of 5,000 CVs on the road by 2018. • Focus on five V2I applications: reduced-speed work zone warning, curve warning, spot weather warning, red light violation waring, and pavement management system. • Expansion underway for 40 additional locations in Ann Arbor; deploying on 350 miles of roadway by 2019.
Ali Zaghari, Caltrans	**I-710 Truck Lanes** • Caltrans is constructing a two-lane, truck-specific facility for trucks leaving the Port of Long Beach along I-710. • The facility aims to relieve congestion caused by excessive trucks from the port. • This controlled access facility could be equipped with CV–AV applications.

FACILITATED BREAKOUT GROUP DISCUSSION

Key Points from Freeway and Managed Lane Performance Discussion

- The addition of CV–AVs could cause problems related to unacceptable degradation of the system.
- There is no desire from an operator standpoint to make a CV–AV lane until there is a higher market penetration.
- Is there enough benefit from CV–AV applications such as CACC to increase capacity to allow CV–AVs?
- There is no CV–AV allowed class for managed lanes, and the faculties need to serve HOV and motorcycles first.

Key Points from Infrastructure and Implementation Discussion

- There is a need for supporting infrastructure and vehicle technology.
- There needs to be guidance on what striping and signage is required by the system.
- There needs to be a state-of-the-art TMC for hosting and running the applications in order to ensure functionality and support the system.
- There is a need for accurate mapping of the system, including knowledge of the access control.

Key Points from Design Guidelines Discussion

- There is soon to be a V2I guideline released.
- There needs to be tools for developers to use to get started.
- A system to track configuration and manage ITS (including CVs) is needed.
- Steps to identify and address a failure in the CV system quickly must exist.

KEY TAKE-AWAYS AND LESSONS LEARNED

Infrastructure–Implementation

- Key unknowns: CV–AV effect on level of service, changes in managed lanes eligibility for CV–AVs, changes in federal policy for CV–AVs.
- Need to communicate benefits and impacts to stakeholders and public.
- Assess how CV–AVs can increase capacity or decrease level-of-service.
- A sufficient model or tool to evaluate CV–AVs on managed lanes does not exist.
- Must determine what needs, such as mapping, IT support, and V2I applications exist for systems to function.
- Need for mapping of facility and access points.
- International technology transfer and globalization of applications would enable better operations.

Operational Guidelines

- Key unknowns: CV–AV penetration for first installation, staffing resource needs, and how to operate with mixed traffic.
- Federal policy may evolve to give better instruction for CV–AVs on managed lanes.
- Could have different pricing for CV–AVs.

MOVING FORWARD

- Continue the dialogue.
- Next year: session on CV–AV Modeling and Tools for Freeways and Managed Lanes to determine status of current and needs for new tools.
- Next generation transportation management system that utilizes CV–AVs must be developed.
- What does the transition to dedicated lanes for CV–AVs look like?

APPENDIX N: BREAKOUT SESSION 14

Reducing Conflicts Between Vulnerable Road Users and Automated Vehicles

Organizers and Reporters

- Ryan Greene-Roesel, San Francisco County Transportation Authority; and
- Justin Owens, VTTI.

SESSION FOCUS

This session focused on the interaction of AVs and VRUs, including pedestrians and bicyclists. Current research activities were presented and participants discussed issues and opportunities associated with AVs and VRUs. The PowerPoint presentations used by speakers in the session are available on the AUVSI website at http://www.automatedvehiclessymposium.org/home.

SESSION SUMMARY

The session featured the following four presentations.

What Is the Current State of Pedestrian–Vehicle Interaction?

Justin Owens, *Center for Vulnerable Road User Safety–VTTI*

- Owens discussed using current driver and VRU interactions to highlight areas where technology may pose particular challenges and where technology could improve the current state of interactions. He reviewed current VRU crash data. He described factors impacting interactions including visibility, choice of crossing location, choice of when to cross, social factors, and behavioral interplay between drivers and VRUs.
- Owens identified possible opportunities to reduce vehicle and VRU crashes with automation, advanced sensing, and V2X application. He noted that one challenge was not to overly burden VRUs in the process.

Vulnerable Road Users in the Age of Automated Vehicles: How to Ensure Safe Interactions?

Azra Habibovic, *Viktoria Swedish Information Communication Technology Institute*

- Habibovic discussed the current interaction between drivers and pedestrians and how this interaction might change with AVs. He noted that understanding each other's intent through active signals and passive signals, as well as driving and walking behavior is crucial for safe and pleasant interactions. He described some of the research efforts underway and outlined possible issue associated with studying the interaction of AVs and VRUs.

- Habibovic summarized one observational field experiment involving 13 pedestrians experiencing four different driver behaviors : the driver making eye contact, the driver talking on a cellphone, the driver reading a newspaper, and no driver at all. A second test examined pedestrians' emotional reaction to photographs of drivers making eye contact, looking straight ahead, talking on a cellphone, reading a newspaper, and sleeping. Pedestrians' emotional experiences reflected calm interaction with the eye contact alternatives, but stressful and unpleasant interaction with distracted drivers and no driver.

Connected Pedestrians at Signalized Intersections in a CAV Environment

Larry Head, *Systems and Industrial Engineering–University of Arizona*

- Head discussed the need for traffic signals in the future with AVs and the interaction with pedestrians at intersections if there are no signals. He reviewed the U.S. DOT Connected Vehicle Pooled Fund project: MMITSS.
- Head described the Mobile Accessible Pedestrian Signal System, which is an element of MMITSS. He discussed the connected pedestrians and infrastructure systems and the basic concept of operations plan. The system includes a smart phone app allowing a pedestrian to request a crossing signal and provides the count-down time in the pedestrian signal phase. He noted some of the challenges in estimating pedestrian locations.

PROSPECT Project Overview

Andres Apracio, *Product Manager–Applus IDIADA*

- Apracio described current crashes involving vehicles and VRUs and the potential for AV safety systems to reduce these crashes. He outlined the results of an EU-sponsored study examining forward-looking integrated pedestrian safety systems. He also reviewed existing systems.
- Apracio discussed PROSPECT, which is a EU-sponsored project focusing on significantly improving the effectiveness of active VRU safety systems by expanding the scope of scenarios addressed by the systems and improving overall performance. The project includes nine EU countries and 17 partners : five vehicle manufacturers, three suppliers, five research centers, and four universities. Key aspects of the project include better understanding of relevant VRU scenarios, improving VRU sensing, developing advanced system control strategies, validating the system, and conducting demonstrations.

MOVING FORWARD

- Participants discussed that some groups have suggested that traffic signals will not be needed in the future with AVs, but it is unclear how those intersections would safely accommodate VRUs. More research on this topic would be beneficial.

- Participants discussed that OEMs and researchers are developing different approaches for communicating a vehicle's intent to VRUs and other road users. Participants noted that it was not clear if work is being done or should be done to develop methods for VRUs to clearly signal their intent to a vehicle. Participants noted that additional research on this topic would be beneficial.
- Participants suggested that continuing this discussion at the 2017 AVS would be beneficial.

APPENDIX O: BREAKOUT SESSION 15

Behavioral Experiments for Modeling Adoption and Use of Autonomous Vehicles

Organizers

- Yoram Shiftan, Technion University, Israel;
- Joan Walker, University of California, Berkeley;
- Dimitris Milakias, TU Delft; and
- Srinivasan Sivaramakrishnan, University of Florida.

Reporters

- Yoram Shiftan, Technion University, Israel; and
- Katherine Turnbull, TTI.

SESSION FOCUS

This breakout session focused on bringing together travel demand and behavioral researchers to provide structure, guidelines, and direction for behavioral experiments to support modeling, planning, and policy for AVs. The focus was on behavioral experiments and gaining behavioral insights.

SESSION SUMMARY

The session included speakers giving ultra-lightning talks on different approaches, including simulation-based scenario analysis studies, stated-preference surveys, virtual reality and game simulators, revealed-preference surveys, and naturalistic experiments. Participants discussed possible impacts of AVs on vehicle ownership and use, travel behavior, mode of travel, activities and lifestyles, and land use. Topics for further research were identified by participants. A PowerPoint presentation including one slide per speaker for the ultra-lightening talks is provided on the AUVSI website at http://www.automatedvehiclessymposium.org/home.

Overview and Typology

Yoram Shiftan reviewed the session objectives described previously and the following questions to help guide the discussion after the presentations. Shiftan and other session organizers discussed background information to help frame the topics covered in the session.

- How to design reliable choice experiments?
- How to deal with the lack of experience with AVs?
- What creative virtual realities, games, and simulators can better reflect the AV world?

- What type of revealed-preference data can be used today to research behavior in a world of driverless vehicles?
- How to design field experiments?
- Other new methods and creative techniques?

Background

- AVs will have far-reaching implications on travel behavior, activity participation, and land use.
- We need to prepare for the arrival of this technology by developing appropriate modeling frameworks.
- Current behavioral models are limited in capturing behavior in the world of AVs.
- There is an outstanding challenge of developing credible experiments to study the behavioral impacts of AVs.

Activity and Possible Travel Behavior Implications

- Reduce driver burden (stress, fatigue, productive time),
- No need to park,
- Travel time budget,
- Travel money budget,
- Longer commute,
- Travel distance to other purposes,
- Changes in activity patterns,
- More travel,
- Residential location,
- Land use,
- City expansion, and
- Value of aggloremation.

Limitations of Traditional Approaches

- Focus on safety and human factors rather than travel behavior (driving simulators, controlled test beds).
- Assume travel behavior implications (microsimulation, network analysis).
- Ask about hypothetical scenarios that are too unfamiliar to the subjects (stated-preference studies).

General Modeling Challenges of Automation

- Changes in the utility of various modes.
- New modes of driverless vehicles.
- Substitution between modes.
- Changes in the value of time.
- New range of attributes (cost).

- Change in attitudes and preferences.
- The role of societal and cultural factors.
- The role of control seeking and driving fondness.
- The role of policy.
- The penetration and adoption phase.
- How to represent the different automation phases?

Lack of Knowledge and Experience

- Awareness, knowledge, and experience are all important concepts when modeling.
- Adoption of any kind of new technology.
- How do we avoid behavioral bias when trying to measure adoption intentions and when trying to model the different dimensions of behavior implications?

Do We Need a New Integrated Framework?

- System impacts of full automation.
- Fully integrated model to take into account the interrelationships between various choice dimensions: land use, travel demand, traffic microsimulation:
  - Residential location,
  - Activity-based travel demand, and
  - Type of operation (sending the car back or pay for parking).
- Issues: multidimensional.

Data Collection for Analog Modes

- Behavioral response, modality styles, diffusion, adoption, and network effects,
- Car-sharing services (ownership–membership),
- On-demand services (multitasking–value of time), and
- Electric cars (energy efficiency–new technology).

Typology of Research Objectives

- Ownership and use:
  - Willingness-to-pay and preferences,
  - CAV vehicle technology, and
  - Sharing (including versus ownership) acceptance/reservation/service types.
- Travel behavior and mode:
  - Value of time and multitasking,
  - Travel time,
  - Waiting for shared ride, and
  - Modal shift.
- Travel behavior activity and lifestyle:
  - Reschedule activities,
  - Time use, and

  - VMT.
- Land use:
  - Residential location and
  - Work location.

ULTRA-LIGHTNING TALKS

Activity Based Models and Scenarios

- Arun Kuppam, Cambridge Systematics, discussed research forecasting on futuristic travel behavior resulting from AV usage. The project examined how to estimate the impacts of AVs on travel choices and how to use the results to aid in the decision-making processes. The method tested scenarios and ranges of variables in a series of activity-based model (ABM) components.
- Martin Milkovits, Cambridge Systematics, described an Florida DOT project exploring emerging technology, demographic changes, and travel behavior. The project examined paradigm shifts that may influence travel forecasts by developing and testing scenarios using the Southeast Florida ABM to explore model sensitivities and identify areas of data need. Five independent scenarios and five combined scenarios were defined. For the AV scenarios, preliminary results show an increase in average tour length, but a reduction in the longest trips, and an increase in the number of stops per tour.

Stated-Preference Surveys

- Francesco Ciari, ETH Zurich, described a mode choice experiment with shared and pooled AVs. The project is exploring willingness to pay and the market potential for shared and pooled AVs, the influence of public transport accessibility, and the value of time compared to other modes of transportation. The methodology includes a stated preference survey for mode choices in the metropolitan area of Zurich, Switzerland.
- Eric Molin, TU Delft, described a study examining the long-term spatial and transport impacts of AD. The project is exploring the second order, longer-term effects of AD on mobility and spatial behavior and the potential to counteract positive short-term effects. The methodology includes a series of stated choice experiments, attitude and perception measurements, and discrete-choice modeling.
- Ricardo Daziano, Cornell University, discussed a project examining consumers' willingness to pay for AV features. The methodology uses discrete-choice experiments.
- Kara Kockelman, University of Texas at Austin, described a project focused on Americans' preferences and plans for AD technologies. The project includes conducting three online surveys to obtain information on a household's current vehicle inventory and future vehicle purchases, willingness to pay for various vehicle technologies, travel characteristics, and demographics. One of the surveys is being conducted in Austin, one in another area in Texas, and one in the United States.
- Stefan Trommer, German Aerospace Center, described a project examining the impact on value of time, mode choice, and VMT for perspective users of private autonomous vehicles, and autonomous vehicles on demand for different trip purposes, including work,

leisure, shopping, holiday, and business trips. The methodology includes a survey of approximately 1,000 automobile drivers featuring a stated-preferences experiment. The survey results will be incorporated into a microsimulation model to calculate the impact on mode choice and VMT for the region of Brunswick, Germany.

- Yoram Shiftan, Technion, described a study examining user preference regarding AVs The project focuses on identifying factors influencing the decision to buy, subscribe, and use private and shared AVs versus current automobile ownership. The project is also examining the difference in AV preferences for vehicle type between Israel and North America and the willingness to pay for private and shared automation. Stated choice experiments, attitude, and perception measurements, and discrete-choice analysis are being used in the study.

Stated Preference, Virtual Reality, Gaming, and Simulators

- Simons Washington, Queensland University of Technology, described a study examining the willingness to pay for future mobility options focusing on consumers in Australia, Indonesia, Malaysia, and Thailand. The study is exploring what mobility options are consumers willing to pay for in 2030 and how car-sharing, automation, and alternative fuels are being considered. The study methodology includes stated preference experiments, revealed-preference demographics and mobility investments, a 1-day travel survey, and consideration of future options using scenarios and video.
- Bilal Farooq, Polytechnique Montréal, described immersive virtual reality experiments focusing on AV ownership, sharing, and use. Agent-based traffic microsimulation and simulation techniques are being used in the project.
- Matthew Beck, University of Sydney, discussed research examining how drivers engage with autonomy and the effect of experience on attitudes and preferences. The study methodology includes the use of network connected driving simulators and experimental economic and stated preference surveys.
- Azra Habibovic, Viktoria Swedish ICT, discussed a project called, Wizard of Oz: A Technique for Studying Interactions with Automated Vehicles. The study is examining if tactical control is desirable while driving in an automated mode. The methodology includes using a driving simulator to simulate highly automated driving. Each test driver experienced two of three available tactical controllers.
- Hani S. Mahmassani, Northwestern University, discussed Pokemon-Go and Autonomous Cars. The game places little (virtual) Pokemons in various locations. To battle Pokemon, a player must be at a close physical distance. A player can choose how to chase and collect Pokemons, including walking, bicycling, driving, and using a robot car. The project aim is to observe how the player utilizes autonomous vehicle capabilities to accomplish both personal and household members' activities.

Revealed Preference

- Chris Schwarz, University of Iowa, discussed a study examining the transfer of control from HAD focusing on performance and trust using a driving simulator. The project included 20 participants driving one practice and two main drives with Level 3 automation, while engaged in a trivia task. One drive had take-over requests at problem spots, such as work zones, and the other drive automatically responded by changing lanes and other maneuvers.

- Giovanni Circella, Georgia Institute of Technology, described a project examining the impact of travel multitasking on mode choice and the implications for AVs. The project focused on the question how do the activities conducted while traveling affect the utility of public transportation today and AVs in the future. The project used a mode choice model estimated with behavioral and attitudinal data collected in northern California.
- Joan Walker, University of California, Berkeley, described a naturalistic experiment with chauffeurs to explore how people would use their automobiles differently if they were fully autonomous. The naturalistic experiment provided automobile-owning households with 40 h of chauffeur service. Travel with and without chauffeur is being tracked using a smartphone.

DISCUSSION

- Participants discussed that individual behavior is key to the impact AVs will have on the transportation system, land use, and human health. Depending on individual behavior, the benefits may be positive, negative, or mixed.
- Participants discussed the advantages and limitations of different research methods. It was suggested that multiple approaches are needed to better understand the potential impacts of AVs on traffic congestion, mobility, land use, and the environment.

MOVING FORWARD

- Participants suggested that an integrated approach using multiple methods would provide the most robust data to address key questions. It was further suggested that additional research was needed to outline the key elements of this integrated approach.
- Participants suggested that developing better methods to provide experience and knowledge of AVs to respondents would helpful. It was also suggested that research developing these methods would be beneficial.
- Participants discussed that individual's knowledge, awareness, and preferences will change over time. As a result, participants identified the need for research and experiments to continue over time and to collect consistent longitudinal data and data across geographies.
- Participants discussed how coordination and collaboration with other research activities, including leveraging field operational tests for behavioral research would be beneficial. It was suggested that including research in travel, activity, attitude, and behavioral changes in all field operational tests and pilots would also be beneficial.
- Participants discussed how research that could develop a standard set of before-and-after questions to use in all experiments would be helpful.

APPENDIX P: BREAKOUT SESSION 16

Aftermarket Systems (ADAS-Related)

Organizers

- Chris Borroni-Bird, Qualcomm Technologies Inc.;
- Jim Misener, Qualcomm Technologies Inc.; and
- Shawn Kimmel, Booz Allen Hamilton.

Reporter

- Katherine Turnbull, TTI.

SESSION FOCUS

This session focused on obtaining a better understanding of the role that aftermarket systems may play in accelerating the deployment of AVs. The presentations addressed ADAS and other after-market systems that provide safety benefits and other driving enhancements. The session brought together speakers from Silicon Valley's start-up companies, established Tier 1 suppliers, and commercial fleet operators to discuss the benefits and challenges associated with aftermarket system deployment and potential business models.

SESSION SUMMARY

The session featured the following seven presentations and a panel composed of the speakers. Copies of the PowerPoint presentations used by some of the speakers are available at http://www.automatedvehiclessymposium.org/home.

Return of the Aftermarket

Roger Lanctot, *Strategy Analytics*

This presentation examined the recent growth, current status, and potential future expansion in the ADAS market. Suggested factors supporting more widespread use of ADAS included lower system costs, public programs to reduce road fatalities and improve safety, automakers continued need for new features, and growing consumer interest. In addition, it was noted that ADAS are part of the evolutionary path to AVs. The use of different technologies, approaches, and systems were highlighted.

Aftermarket Truck Platooning and Automation

Steve Boyd, *Peloton Technology, Inc.*

This presentation described the use of ADAS in the trucking industry, including truck platooning. Recent examples of truck platooning pilots and demonstrations were highlighted. Peloton system and activities under development were described.

Utility, Economic, and Business Models for Aftermarket ADAS Today and in the Near Future

Chris Brogan, *AssureNet*

This presentation focused on insurance business models supported by aftermarket ADAS. Advantages from ADAS include reduced losses and increased efficiencies. Further, it was noted that when combined with ride event recording data, ADAS can streamline claims management and reduce loss costs. The potential savings from different systems were summarized.

Driving Behavior Detection Technology for ADAS–AD and Its Application for Smarter Mobility

Hiro Onoda, *SVP Denso International America, Inc.*

This presentation focused on ADAS and driving behavior detection technology. Technologies that can be used to detect driving behavior were described, along with developing systems to detect and respond to bad or wrong behaviors.

Future of Urban Mobility Is Not Autonomy— It's Autonomous, Connected, Electrified, Shared

Stefan Heck, *Nauto*

This presentation focused on the use of ADAS in combination with data on good and bad driving behaviors to develop harmonious autonomous vehicles. It also highlighted other multimodal and share-fleet needs in urban areas and using autonomy to enhance livability in cities.

Fleet Operator's Perspective on Retrofitting Aftermarket ADAS Technologies

Scott Perry, *Ryder Fleet Management Solutions*

This presentation examined the potential benefits of truck platooning from a fleet operator's perspective. Factors contributing to the truck driver shortage were described. The evolving role of technology in the trucking industry was reviewed, along with considerations for the use with specific fleets.

Role of V2X-Based Aftermarket Safety Devices in Smart Cities

Paul Sakamoto, *Savari*

This presentation examined the use of DSRC in V2V, V2I, and vehicle-to-pedestrian (V2P) applications. Examples of current and potential approaches were highlighted, including the use of aftermarket safety devices.

MOVING FORWARD

- Participants suggested that developing a catalog of the diverse aftermarket opportunities would be beneficial. The catalog could be categorized by technology, such as DSRC-based and camera-based systems, and by mode, fleet, and user.
- Participants suggested that identifying the progression of each of these opportunities to different levels of automation would also be beneficial.

APPENDIX Q: BREAKOUT SESSION 17

Policy Making for Automated Vehicles
A Proactive Approach for Government

Organizers

- Anita Kim, Volpe Center, U.S. DOT;
- Ginger Goodin, TTI;
- Steve Lockwood, Lockwood LLC;
- Baruch Feigenbaum, Reason Foundation;
- Richard Mudge, Compass Transportation and Technology;
- Shawn Kimmel, Booz Allen Hamilton;
- Natarajan Jana Janarthanan, Washington State DOT;
- David Perlman, Volpe Center, U.S. DOT;
- Joe Stanford, Volpe Center, U.S. DOT;
- Susan Spencer, Susan Spencer & Associates;
- Casey Emoto, Valley Transportation Authority; and
- Chuck Fuhs, TRB Managed Lanes Committee.

Reporters

- Ginger Goodin, TTI;
- David Perlman, Volpe Center, U.S. DOT; and
- Anita Kim, Volpe Center, U.S. DOT.

SESSION FOCUS

This session focused on some of the key automation policy priorities facing public agencies and the strategies for policymakers to achieve desired outcomes. The session explored a range of automation scenarios to understand the broader policy implications. Panel presentations and interactive discussions were used to highlight different policy perspectives and strategies. Policy actions that enable public agencies to encourage the benefits from AVs, while avoiding potentially negative impacts, were discussed.

SESSION SUMMARY

Opening Remarks

Anita Kim, *Volpe Center*

- Kim noted that the public policy landscape is broad and diverse, and encompasses all levels of government and agency leadership. She reviewed the session objectives to engage a breadth of perspectives, both public and private, and describe a range of policy strategies, particularly for state and local governments. Of particular interest is how AVs can support policy goals through various levers, not only in the near term but for the longer term as well.
- To set the stage for the discussion, Kim noted that there are a number of key challenges that policy makers are called upon to address.
  - Avoiding the patchwork of state regulations, and the need for consistency.
  - Understanding legal and liability issues.
  - Data privacy: ensuring appropriate data management practices.
  - Driver licensing and training.
  - Benefits versus risks and unintended consequences.
  - Broader societal implications of AVs and equity issues.
  - Infrastructure requirements for enabling AVs.

Panel 1: Policy Perspectives from Industry Stakeholders

Shawn Kimmel, *Booz-Allen Hamilton, facilitator*

Speakers	Summary of Panel Discussion
Paul Scullion, Association of Global Automakers, Inc.	• Scullion noted that it was important to understand how technology and policy interact, and how public benefits can be achieved. It is complex and requires continuous engagement and education. • He reviewed some of the efforts in the past year, including 35 legislative proposals at the state level and the NHTSA interpretation to Google on defining a driver. • Understanding driver behavior, and re-engagement when needed, will be an important issue. One topic to consider is impaired driving laws and related policy, and the potential legality issues regarding impairment at lower levels of AV.

Panel 1: Policy Perspectives from Industry Stakeholders (*continued*)

Shawn Kimmel, *Booz-Allen Hamilton, facilitator*

Speakers	Summary of Panel Discussion
Jill Ingrassia, American Automobile Association (AAA)	• Ingrassia noted that AAA is very interested in safety benefits, mobility, and quality of life. AAA was a pioneer in driver education and that will continue to be a strong interest. • From an industry perspective, everything related to TNCs and MoD concepts feeds into new opportunities. While developments with AVs alone present opportunities, these innovations feed off each other and create new dynamics and partnerships that lead to new thinking about the future of transportation. • Technology is still too much of a moving target to define the necessary steps. AAA often feels confident to advocate based on data. State and local representatives need to ask the right questions to inform the next steps they should take as developments progress. • When decisions are made to be part of pilot tests, agencies need to ensure safety. Part of pilots is also educating and informing the public. It is too early to finalize what driver training for AVs should be. An inventory of the key questions to answer at some point and where to seek answers would be beneficial. • The consumer perspective is important; policy makers need to consider the value proposition to consumers for AVs, which will vary person-to-person and locality-to-locality. That value proposition should consider whether the technology is more reliable, cleaner, and safer—an improvement over the current situation.
Robbie Diamond, Securing America's Future Energy (SAFE)	• Diamond noted that SAFE is focused on ending oil dependence. The organization sees AVs as an accelerator in solving oil dependence. • There is a high level of public awareness of AVs, but also lack of information on the technical details of AVs and energy issues. • He suggested a "get out of the way" approach to regulation, but preemption will be needed to avoid the patchwork problem and to be responsible. Learn by doing rather than legislating; see how people interact with the technology and then make decisions. Regulators should not try to predetermine outcomes. • Expect VMT to increase based on modeling. People may use AV in ways that cannot be anticipated.

Panel 1: Policy Perspectives from Industry Stakeholders (*continued*) **Shawn Kimmel**, *Booz-Allen Hamilton, facilitator*	
Speakers	**Summary of Panel Discussion**
David Strickland, Counsel, Self-Driving Coalition for Safer Streets	• Strickland spoke on behalf of Self-Driving Coalition for Safer Streets, supported by Google, Volvo, Lyft, Uber, and Ford. The coalition is interested in how to test and deploy SAE Level 4 and 5 vehicles as soon as possible, managing human error and creating consumer pull. The coalition is also interested in reducing the possibility of regulatory patchwork. • The U. S. DOT and NHTSA are now dealing with a regulatory process that has been in place for decades, and which is now being challenged by the speed of technology. That is a new normal for any public health and safety agency; it is transformational. • States and localities have an active and needed role now, and there is growing recognition that localities and states are part of the process. For example, there is clear role for safe testing and creating environments for development. There is "on-the-ground" work on items that AVs need, such as better line painting. • The expansion of states and local governments into newer spaces will require expertise from NHTSA. • Data is also important. States are originators of data that NHTSA uses to document fatalities and crashes. The data systems in each state can be improved to increase public safety. This involves not only open availability but also the crash data collection process. • There is a void in the federal regulatory framework. There are AVs on the road now, and the U.S. DOT is working hard to fill that void as quickly as possible. There are laws and regulations in place that are serving the rollout of AVs. There has never been a FMVSS for ADAS, but the OEMs have been willing to take the risk and NHTSA has used defect authority to take action. • There is a larger philosophical issue for NHTSA, which is "foreseeable use." When considering a consumer product, how do you deal with the philosophical nature of making a product that invites experimentation and unintended use that should have been foreseen? • One area that will require some initial investment is city planning. Policy strategies involving land use, transit, last-mile service, road user allocation, congestion pricing, incentives, and parking are needed. Decisions made today can make a city a place to embrace self-driving vehicles, and also generate economic advantages. Cities should consider policies that can enable everyone to take advantage of the societal benefits of AVs. • Policy making should consider safety benefits at every level, and the need to democratize safety. When 35,000 people die in crashes every year, saying people are pretty good at driving is wrong. Half of the people who die are unbelted; a third are impaired. There's a place for higher levels of AVs to alleviate that burden.

Panel 2: Real Policy Challenges from Real Agencies **Baruch Feigenbaum**, *Reason Foundation, facilitator*	
Speakers	**Summary of Panel Discussion**
Bernard Soriano, California Department of Motor Vehicles (DMV)	• Soriano noted that the California DMV was tasked with developing regulations to allow AVs on roads in the state. Regulations for testing AVs have been developed and operational guidelines–regulations to allow the general public to operate AVs are being developed. • The California DMV has been working on these issues since 2013. Input was obtained from NHTSA and other state agencies affected by AV enforcement, insurance, roadways, and other areas. A steering committee was formed to identify and vet issues. • Driver education is not clearly understood in thinking about the operation of AVs. • A focus has also been on how to ensure that AVs are safe. Identifying design elements, thresholds, and processes needed to ensure that vehicles are safe is part of the process. • State regulations addressing the safety features in vehicles did not exist. Stakeholders emphasized the need to experience AV technology, which required developing relationships with automakers. Meetings and technology demonstrations were needed. • The California law requires the DMV to ensure that the vehicles are safe. Answering the question "how safe is safe enough?" is difficult. • Education and outreach programs need to be part of an integrated strategic plan for AVs. • The initial focus is on passenger vehicles. Commercial and other vehicles will be examined next.

Panel 2: Real Policy Challenges from Real Agencies (*continued*) **Baruch Feigenbaum**, *Reason Foundation, facilitator*	
Speakers	**Summary of Panel Discussion**
Tracy Larkin-Thomason, Nevada DOT	• Larkin-Thomason noted that the Nevada DOT became more involved in AVs as infrastructure compatibility questions were increasingly being raised. This topic reaches many aspects of state government service providers. • Data integration between vehicles and infrastructure is of interest to the department. Demonstrations around the Las Vegas Convention Center have been conducted with communication devices on signals. As more requests for other tests are received, it is important to ensure standardization. There have been some compatibility challenges, primarily with the data interface. • Standardization with the infrastructure is also important. Lane striping is refreshed for demonstrations, but it is not possible to maintain that level of service on an ongoing basis. Maintaining the reflectivity of markings in an extreme environment, including heat and snow, is not always possible. • The viewpoint on regulation differs among states. The DMVs in California and Nevada are working together examining safety-focused testing regulations, requirements for insurance, and previous testing experience. • Regulations in Nevada are not as stringent as those in California. In some states, if there are no regulations then the interpretation is that AV testing is not legal. In others, the interpretation is that testing is legal. There is a need for standardization. • The metropolitan planning organization (MPO) in the Las Vegas region has promoted public education in southern Nevada. It included sessions with commissioners and city council members. • Nevada began by focusing on regulations with passenger cars, but the process is expanding to include commercial vehicles. Consideration is also being given to shared AVs. • Balancing public safety while allowing innovation is challenging. The California legislature has provided direction on ensuring safety. Who defines what is safe enough? There are differing opinions based on state culture. There are at least 15 states actively engaged in this area.

Panel 2: Real Policy Challenges from Real Agencies (*continued*) **Baruch Feigenbaum**, *Reason Foundation, facilitator*	
Speakers	**Summary of Panel Discussion**
Nathaniel Beuse, NHTSA	• Buese described his responsibilities at NHTSA, including supporting regulatory work associated with AVs and CVs. The focus is on developing guidelines and model state policies, as well as new analysis tools. • There is a role for states and cities in encouraging safe development and deployment of the technology, recognizing that new approaches may be needed. Typically, testing has been done on test tracks, but AVs need real-world testing and simulation. It is important to engage cities, as AVs are being tested on local streets. • Consumers have to experience the new technologies to make informed decisions. Explaining how autopilot works on Tesla and Mercedes vehicles, which are different, is not the same as experiencing the differences. Exposing the public to the technology is needed to promote understanding. • An important point on the previous discussion on modeling is that the average vehicle age is about 12 years. It takes 25 years for single technology to penetrate the vehicle fleet. Predictions of having pervasive AVs by 2040 does not fit with the vehicle turnover experience. • Some states and cities are considering legislation and policies related to testing and operating AVs. Cities and states often want to be seen as leaders and early adopters, but care should be taken not to impose a patchwork of regulations that would inhibit innovation. A useful exercise for states and localities is to first examine their own regulations for impediments.
Karla Taylor, City of Austin, Texas	• Austin was second city after Mountain View, California, to have Google testing its self-driving vehicle. Austin is known as an early technology adopter city. The U.S. DOT's Smart City Challenge helped make Austin and the region aware of AVs. • There has not been a focus on AV at the city level due to the priority of dealing with the day-to-day needs. There is a realization that AVs are coming and cities need to prepare. There may be mobility and accessibility benefits from AVs. • There is a deficit of knowledge about AVs among the public at large. Meetings, seminars, and other activities are needed to help inform the public and policy makers. It is incumbent on industry and federal agencies to get more proactive in sharing the knowledge. Cities also have to become more knowledgeable. • While policies can be developed, it really comes down to implementation. Everybody's for compact and connected transportation, but "not in my neighborhood." To avoid a nightmare scenario of more traffic congestion with AVs, policies related to encouraging shared use may be needed.

Panel 2: Real Policy Challenges from Real Agencies (*continued*) **Baruch Feigenbaum**, *Reason Foundation, facilitator*	
Speakers	**Summary of Panel Discussion**
Karla Taylor, City of Austin, Texas (continued)	• The Austin central business district has no parking minimums, but those who finance developments require building developers to provide parking, so the business community must also evolve. • Texas does not have regulations at state level that prohibits AVs. • The most important vehicle ownership variables to measure include adoption, cost, and automated features of interest.
Keith Jasper, Northern Virginia Transportation Authority (NVTA)	• Jasper described NVTA, which is an agency created in 2002 to address traffic congestion in northern Virginia. Over past 3 years, the NVTA has allocated close to $1 billion to transportation improvements in a four-county region. • The NVTA is focusing on the smartest way to invest available funding, including examining the potential benefits of AVs. • The NVTA is a good example of a public agency that is still learning about AVs and related technology. • As survey was conducted of residents in the four counties on their perception of when AVs would appear in northern Virginia. Approximately 80% of respondents thought AVs would be ubiquitous within 20 years. Approximately half said they would likely use an AV, while 17% responded they would never use an AV. Understanding the public perception of AVs is important. The survey will be repeated again. There also appears to be political interest in AVs. • As a regional planning and programming organization, the NVTA does not own or operate the roads. If striping is the issue, the NVTA cannot help with that. • The NVTA is conducting long-range planning scenario analyses, which include AVs to complement the base forecasts.
Mike Alexander, Atlanta Regional Commission (ARC)	• Alexander described the responsibilities of ARC, which is the MPO for the Atlanta metropolitan area. The ARC is responsible for long-range planning. The ARC has an activity-based model, and transportation modeling is conducted in concert with land use modelling. It is important to understand the interactions between transportation and land use. • Traffic congestion is an issue in Atlanta. It is one of the driving forces behind the desire to understand the impacts of AVs. There are 133 local governments in the ARC region. Elected officials have an interest in this topic. • The ARC has a "zero death" policy. San Francisco and other communities also have a "zero death" policy. It guides the overall approach to investments and operations.

Panel 2: Real Policy Challenges from Real Agencies (*continued*) **Baruch Feigenbaum**, *Reason Foundation, facilitator*	
Speakers	**Summary of Panel Discussion**
Mike Alexander, ARC (continued)	• The Smart City Challenge was an ingenious approach. The ARC reviewed all the finalists' proposals. Columbus staff are coming to Atlanta to talk and partnerships may be possible. Columbus had a $90 million commitment from the private sector, while Atlanta had none. There is obviously room for Atlanta to improve. • On public education, it is important to remember that many people do not read the owner's manual in their vehicle. • The ARC has modeled the possible impacts of scenarios including AVs. A 50% increase in capacity, with no other improvements, results in a 50% reduction in delay. Trip lengths increase. People drive more but they experience fewer hours of delay. Transit trips decrease. On the policy side, consideration is being given to the integration of mobility options. One example is the use of TNCs for first- and last-mile to support transit. • There is a need to have conversations about what should be built to support AVs. Those conversations have begun in Atlanta. • Meetings have been held with law enforcement personnel to discuss who they arrest when they impound a driverless vehicle. It is a question of how rules will need to change.

INTERACTIVE SESSION: THE FUTURE IS A CHOICE: POLICY LEVERS FOR STATE AND LOCAL AGENCIES

Facilitators: Ginger Goodin, Texas A&M Transportation Institute, and Richard Mudge, Compass Transportation and Technology

This interactive discussion centered on the range of policy levers available to state and local agencies to reach desired outcomes with respect to AVs. As AVs become more widely deployed on public roads, what policy tools can state and local agencies use to mitigate potentially negative impacts and incentivize positive ones?

Approximately 60 attendees participated in the interactive discussion of potential policy levers and joined an informal voting process. A list of policy levers (provided in the following) was adapted by the session planning committee using in-progress research for NCHRP 20-102(01): Policy and Planning Actions to Internalize Societal Impacts of CV and AV Systems into Market Decisions is available at http://apps.trb.org/cmsfeed/TRBNetProjectDisplay.asp?ProjectID=3934.

Automated Vehicles: Possible Policy Levers

- Enact legislation to stimulate AV testing through either legalization of testing or by funding testing activities.

- Establish, codify, and enforce AV operator/owner/passenger requirements, including operator training and licensing, to promote safe operation.
- Accelerate AV market penetration by subsidizing equipped vehicles; both original equipment and aftermarket retrofit of conventional vehicles.
- Increase public and stakeholder awareness through education, communication and outreach, to stimulate consumer action, supportive public investment, and conducive political environment.
- Restructure liability regimes, including insurance requirements, to accelerate market penetration.
- Subsidize shared vehicle services to support ridesharing and transit, including paratransit, to minimize growth in travel demand or maximize accessibility.
- Create economic incentives, such as pretax transit benefits, to support market penetration of shared AVs near transit nodes, urban centers, and commercial centers.
- Implement land use regulations and parking requirements to increase development density in support of market penetration of shared AVs at transit nodes, urban centers, and commercial centers.
- Grant AVs—including transit and commercial vehicles—privileged access (HOV, managed lanes, signal priority, parking access) to encourage adoption.
- Apply road pricing, including tolling, parking pricing, and emerging applications of distance-based pricing, to minimize growth in travel demand or maximize accessibility.
- Increase public agency knowledge and capabilities to capitalize on widespread deployment of AVs and maximize societal benefits.

Connected Vehicles: Possible Policy Levers

- Enact legislation to stimulate CV testing by funding testing activities.
- Accelerate CV market penetration by subsidizing equipped vehicles—both original equipment and aftermarket retrofit of conventional vehicles.
- Invest in CV infrastructure in collaboration with private sector to accelerate V2I deployment.
- Implement new contractual mechanisms with private service providers, including shared data arrangements, to advance connectivity.

Results of Informal Audience Polling

From the 16 potential policy levers, which encompassed economic, regulatory and planning actions by state and local governments, the two key levers identified by participants for action now and over the next 3 to 5 years were

- Increase public and stakeholder awareness through education and outreach to stimulate consumer action, supportive public investment, and conducive political environment.
- Increase public agency knowledge and capabilities to capitalize on deployment of AVs.

There was also strong support for investing in CV infrastructure in the near term, including common data platforms and traffic management systems. There was support in the mid- to longer term for modifying driver licensing, restructuring liability regimes, and looking for opportunities to grant AVs privileged access to lanes or parking.

Actions that involve government subsidies for equipping vehicles—be it AVs, CVs or specifically shared AVs—did not rank high in the informal voting.

Some of the key take-aways and lessons learned from the discussion were

- Automated vehicles are increasingly a part of the public conversation;
- Importance of public and public agency education;
- Technology development and adoption paths remain unclear;
- Risks of prematurely regulating or legislating;
- Cities as early adopters and test beds;
- Importance of long-term transportation planning process and policies; and
- Accepting the "new normal."

MOVING FORWARD

- Educate and inform.
- Review existing regulations–laws.
- Focus first on where you have traditionally played a role.
- Engage with your local community.
- Experience the technology directly.
- Consider long-term planning needs and models.
- Share in order to learn.

APPENDIX R: BREAKOUT SESSION 18

Effects of Vehicle Automation on Energy and Carbon Intensity

Organizers and Reporters

- Don MacKenzie, University of Washington; and
- Paul Leiby, Oak Ridge National Laboratory.

SESSION FOCUS

This session considered potential effects of automation on the energy and carbon intensity of vehicle road travel. Along with travel distance and mode choices, which were addressed in other sessions, these are the major determinants of total energy demand and GHG emissions from vehicles. The session centered on this question: Given improved projections of future travel demand and travel patterns in response to CAVs, what will be the system-wide effects on energy demand and climate change? Speakers addressed the general framework for understanding energy and emissions implications of automation, the impacts of automated mobility services, the effects of driving style and controls, and scaling assessments from individual vehicles and localities up to national level impacts. PowerPoint presentations for some of the speakers are available on the AUVSI website at http://www.automatedvehiclessymposium.org/home.

SESSION SUMMARY

The session included eight speakers under four related themes, with active audience participation throughout.

Setting the Framework	
Speakers	**Summary of Presentation and Question and Answer Session**
Caroline Rodier and Matthew Barth, National Center for Sustainable Transportation	**Energy and Emissions Implications of Connected and Automated Vehicles: Results from a National Center for Sustainable Transportation Study** • Impacts depend on CAV technology attributes and use cases, but the overall assessment is complicated by many countervailing system-level effects. The direction and magnitude of net effect is still unknown. • Increased capacity, reduced time cost, new user groups, and deadheading are all expected to increase travel. • Automated taxis may moderate travel, especially if shared. • Operational efficiency can reduce energy demand and emissions: lots of improvements in the 5% to 10% range can add up. • Lots of trade-offs among safety, mobility, and energy goals. **Key Questions** • Will policy preclude deadheading by privately owned vehicles? How large is the motivation for this behavior? Will this be obviated by widespread adoption of shared-use services?

Energy Impacts of Automated Mobility Services	
Speakers	**Summary of Presentation and Question and Answer Session**
Yuche Chen, NREL	**Estimate of Fuel Consumption and GHG Emissions Impact from an Automated Mobility District** • Adapts tools for analyzing a personal rapid transit (PRT) system in a campus environment, applying the analysis to a service provided by CAVs. • By providing a last-mile solution and enabling personal vehicles to be parked at the campus periphery, total GHG emissions (including the home-campus portion) can be reduced up to 5%.
Kara Kockelman, University of Texas at Austin	**Shared Autonomous Electric Vehicles Operations Across the Austin, Texas, Network, with a Focus on Charging Infrastructure Decisions** • Agent-based simulation using Nagel's MATSim. • Warm start-to-site charging stations. • Results highlight a number of tradeoffs in system design: – Number of stations depends on vehicle range (not charge time). – Waiting time depends on passenger/vehicle ratio, range, and recharging time. – On time performance depends on passenger/vehicle ratio, not so much on recharging time. **Key Questions** • What if you allow some portion of the fleet to be non-EVs? What effect might on-line inductive charging in special lane segments have on performance by eliminating recharge stops?

Effects Of Driving Style and Control	
Speakers	**Summary of Presentation and Question and Answer Session**
Jun Liu, University of Texas at Austin	**Anticipating the Emissions Impacts of Autonomous Vehicles Using the MOVES Model** • Used MOVES to simulate the effects on emissions from smoothing common drive cycles, which reduces acceleration and deceleration levels. • Most emissions are reduced, but a few can increase, such as carbon monoxide (CO) and CO_2 on a Urban Dynamometer Driving Schedule cycle. **Comments** • Consider using a modal emissions model such as CMEM or VT Micro for these types of analyses. • There are rich data sets available on real-world driving patterns. These can be better to use rather than to rely on drive cycles.
Dominik Karbowski, Argonne National Laboratory	**Impact of Connectivity and Automation on Advanced Vehicles Energy Use** • Used Autonomie to simulate fuel consumption effects of smoothing acceleration and braking, reducing stops, and increasing average speed. • Fuel savings strongly dependent on powertrain and speed. • Previewed ongoing work to integrate Autonomie with the POLARIS agent-based simulation environment. **Key Questions** • What are benefits of smoothing smaller at higher speeds (there are fewer irregularities to smooth out). – Several questions about POLARIS. It uses a mesoscopic traffic mode with Markov chains to develop naturalistic speed profiles. Very computationally efficient (written in C++). Embeds an activity-based travel demand model.
Matthew Barth and Xuewei Qi, University of California Riverside	**Environmental Impacts of Connected and Automated Vehicles: Research Update** • Presenters gave an overview of modeling tools for simulating the effects of operational changes. • Example of PHEVs: can reduce energy consumption 14% using current information, but up to 70% if you know future drive cycle perfectly. • There can be spillover effects—positive or negative—from eco-control strategies to unconnected cars. **Key Questions** • We talk a lot about how automation might affect behavior. Do you think connectivity alone might also affect behavior?

Scaling Up from Local Results to National Impacts	
Speakers	**Summary of Presentation and Question and Answer Session**
Don MacKenzie, University of Washington	**Scaling of Platooning Energy Benefits with Compatible Vehicle Adoption** • If platooning requires V2V communication, then vehicles must be compatible. Few compatible vehicles means few opportunities to save energy, and so little incentive for consumers to adopt. • Energy savings per vehicle grow quickly with adoption, which suggests policies should incentivize initial adopters.
Jeff Gonder, NREL	**Potential Energy Impacts of Connected and Automated Vehicles: Opportunities and Approaches** • Consortium of labs is considering the prevalence of different driving conditions across the country, and the applicability of various benefits of automation. • Modeling adoption of CAV technologies by different market segments (e.g., integrating into MA3T model). **Key Questions** • What is the right nesting structure for choosing a CAV versus other vehicle choices in nested logit model? • How do you scale this to the whole country, since it's clearly impractical to run hundreds of unique regional travel demand models? • Where do you get the data for regional congestion levels (trying to disaggregate HPMS)? • How are effects combined/compounded?

CONCLUSION

The results of this breakout should be considered along with those of other breakouts, particularly those addressing the adoption of CAV technologies, and travel behavior implications of CAV adoption. The direction and magnitude of the net effect of automation on energy demand and GHG emissions remain unknown. However, there is an emerging consensus that CAVs will probably improve operational energy efficiency, but could very well lead to increases in VMT. A shift toward mobility services in place of private vehicle ownership is seen as a key enabler for key energy-saving opportunities, particularly given shared rides and right vehicle sizing—and for mitigating induced demand. Effects are likely to be heterogeneous across regions and across individuals, and fundamental research is needed at both the micro and systems levels to accurately anticipate and effectively manage climate impacts of automation.

APPENDIX S: BREAKOUT SESSION 19

Cybersecurity and Resilience Challenges and Opportunities for Self-Driving Vehicles

Organizers

- Edward Fok, FHWA;
- Andre Weimerskirch, University of Michigan Transportation Research Institute;
- Alan Chachich, Volpe Center, U.S. DOT;
- Reinhard Pfliegl, Austrian Association for Advanced Propulsion and Automated Driving; and
- Chuck Price, Peloton Technology, Inc.

Reporters

- Edward Fok, FHWA; and
- Katherine Turnbull, TTI.

SESSION FOCUS

This session focused on three major objectives. The first objective was identifying and discussing the security, safety, and resilience goals of self-driving vehicles. The second objective was identifying and discussing the challenges and opportunities the automotive kill chain presents for SDVs. The third objective was identifying and discussing the challenges and opportunities in an integrated automated transportation network. PowerPoint presentations for some of the speakers are available on the AUVSI website at http://www.automatedvehicles symposium.org/home.

SESSION SUMMARY

The session included the following five presentations followed by a discussion of research opportunities.

Attack Scenario on Automated Vehicle Platoon

Ryan Gerdes, *Utah State University*

- Gerdes discussed the vulnerability of AV platoons to potential perturbation to the platoon control parameters. The perturbations may be accidental or caused by tampering. He described how platoon control logic could be weaponized to cause a collision.

Integration of Self Driving Vehicles into the Traffic Management and IoT Ecosystems

Brian Russell, *Leidos*

- Russell noted the importance of conducting research to identify the vulnerabilities of SDVs. He described potential threat actors, treats, and impacts.
- One vulnerability of SDVs is their significant code base that may be open to analyzing and attacking. Additionally, SDVs will not operate in a vacuum and may be vulnerable to attacks on communication interfaces and wireless access points. The potential for rogue apps is a further vulnerability. SDVs may represent attractive targets for hackers, criminals, and terrorists. Infrastructure elements supporting SDVs may also be vulnerable.
- Examples of challenges to securing SDVs include manufacturers using new technologies and integration methods, consumers' interest in connections to nonvehicle services, and multiple protocols for different devices. Recruiting and retaining skilled cybersecurity personnel is another challenge.
- A security engineering approach considers the traffic ecosystem as a system-of-systems, with each component analyzed from a security perspective. Security is also built into system specifications. Additional security controls are needed for all elements.

Security for Connected Automated Driving

Tim Leinmüller, *Denso*

- Leinmüller discussed the development from CVs to connected automated driving (C-AD) and security for C-AD. The moved from CVs to C-AD increases the number of vulnerable links in the system and the need for additional security.
- Security with C-AD is required to cover the entire in-vehicle system, to cover outside vehicle functions, to allow for secure updates for all functions, and to support privacy across all communication technologies used.
- Examples of security approaches for C-AD include security and privacy by design, security life-cycle management, and security as a process. Integrated multilayered approaches focus on protocols, key management, and hardware.
- To address C-AD requirements, vehicle connectivity and architecture will evolve. Security will need to evolve correspondingly from both a process and a technology standpoint.

Assessing the Risk of Cyber Infrastructure Vulnerabilities:
From Legacy to Next Generation

Kenneth Perrine, *University of Texas at Austin*

- Perrine discussed a study exploring what might happen when legacy and next-generation infrastructure in an urban center is subject to a cyberattack. The study examined what can be exploited, how likely is an exploitation, and the extent of damage that can occur. The results can assist in estimating risk, convincing stakeholders of the need for action, prioritizing vulnerabilities, and assigning resources.

- The legacy system analysis focused on a cyber security incident that disabled traffic signals. The analysis used a dynamic traffic assignment and cell transmission model simulation for downtown Austin, Texas. The analysis estimated that the cost of delay to motorists when 26 of the most-used traffic signals were disabled.
- Considering how the existing cyber infrastructure may support future CAVs is needed. Existing vulnerabilities may still be relevant with CAVs.

Truck Platooning

Chuck Price, *Peloton Technologies, Inc.*

- Price described cybersecurity concerns, assumptions, and approaches for automated truck platooning systems. He noted some of the security elements included in the system.

BREAKOUT DISCUSSION

- Speakers and participants discussed that the long service life of trucks and HDVs is a potential issue with deploying truck platooning. The National Motor Freight Traffic Association has a study group examining heavy vehicle security. The potential to learn from military applications was noted by some participants.
- Speakers and participants discussed that the clustering of complex systems can cause unintended vulnerabilities.
- Speakers and participants discussed the opportunities for vehicles and the roadway system to cooperate to improve misbehavior detection and to improve the security of the overall network.
- The assignment of threat actors or the potential attackers were discussed by the speakers and session participants. Threat actors may include kids hacking into the system to change traffic signals, activists protesting a project, criminals, and international terrorist organizations.
- Speakers and participants discussed attack scenarios, including the failure mode in highly complex systems. Platoon telematic attacks, which can cause a collision in a high-speed platoon, were discussed, as were random area attacks against a vehicle system.

MOVING FORWARD

- Research on writing specifications for automated systems that are secure and useable to a wide audience was identified by participants.
- Examining the appropriate balance between security and usability, including the difference between what the consumer expects and the need of the technical–engineering community was another research topic discussed in the group. It was suggested that systems can be made very secure, but if it is too complex, people will not use it.
- Examining methods to enable and measure crowdsourced data, either public or private, to improve misbehavior detection and to improve resiliency was another research topic suggested by participants.

- Developing and maintaining a sufficiently useable list of potential attacks, their severity, and their probability to enable an effective discussion about possible responses was identified by participants as a possible research topic.
- Participants suggested that research developing methods to gauge the effectiveness of specific security or resiliency solutions would be beneficial.
- Participants discussed the importance of an open security or resilience research platform for AVs and if AUTOSAR being be updated.

APPENDIX T: BREAKOUT SESSION 20

Infrastructure, Work Zones, and Digital Infrastructure

Organizers

- Paul Carlson, TTI;
- Carl Andersen, U.S. DOT; and
- Jerry Ullman, TTI.

Reporter

- Keven Balke, TTI.

SESSION FOCUS

There is growing interest related to the connection between the physical highway infrastructure, digital mapping, and autonomous vehicles. The first third of this session featured speakers who shared their research and views on how the physical highway infrastructure may evolve or be managed differently. The middle third of the session moved toward a more specific and dynamic scenario: accommodating work zones. The final portion of the session focused on digital infrastructure and dynamic mapping needs for autonomous vehicles. PowerPoint presentations from many of the presentations are available on the AUVSI website at http://www.automated vehiclessymposium.org/home.

SESSION SUMMARY

The agenda included 12 speaker presentations on the following pages.

Physical Infrastructure **Paul Carlson**, *TTI, moderator*	
Speakers	**Summary of Presentation and Question and Answer Session**
Hideki Hada, Toyota Motor Engineering and Manufacturing North America	**Vehicle–Machine Vision Interaction with Traffic Control Devices** • ADAS system performance is assessed against targets, but the creation of "good" targets is difficult in test environments. "Targets" need to reflect driving conditions to be encountered in the real-world. • V2X communication may be a way to bridge a pathway to better testing. Information from, and about, the road environment can be used to enhance capabilities and performance of automated vehicle control systems. • There is a need for the automotive industry and infrastructure providers to work together to improve both road design and vehicle systems. Inter-industry dialogue is essential for ensuring good performance of vehicles and to avoid conflicts between smart cars and smart roads. • Communication between industry associations may be more efficient than all OEMs trying to talk to all states independently.
Jian Wang, Carnegie Mellon University	**Smartphone-Based Infrastructure Monitoring** • Wang presented a new smartphone application designed to collect information related to pavement surface quality. • The application used a smart camera mounted to the inside of a vehicle to record the image of pavement conditions as the vehicle traverses a roadway. • Video imaging techniques are used to process images to identify and quantify the type and extent of road damage (potholes, cracking, etc.) • The application is used to classify road quality in the Pittsburgh, Pennsylvania, area. • The application could be expanded to monitor the need for snow and slush removal, as well as sign and pavement marking quality.
Paul Carlson, TTI	**Road Markings for Machine Vision: NCHRP Project 20-102(6)** • Carlson discussed this new research project examining how the performance characteristics of pavement markings affect the ability of machine vision systems to recognize them. • Pavement markings with different levels of performance characteristics will be examined under different operating conditions (daytime: dry and wet, high sun; nighttime: dry versus wet, with versus without roadway lighting, and different headlamp technologies). • The ultimate goal is to provide data for agencies to develop guidelines and criteria for deployment.

Physical Infrastructure (*continued*) **Paul Carlson**, *TTI, moderator*	
Speakers	**Summary of Presentation and Question and Answer Session**
Scott Kuznicki, Toxcel, LLC	**AI-Ambiguous Infrastructure: When Signing and Pavement Markings Don't Make Sense to Drivers or Machines** • Consistency in applications of signs and markings is critical for detection algorithms to work properly. • The roadway environment has inconsistencies everywhere. One example – option lane at exits and multi-lanes exits. Technology needs to "learn" all potential applications of different arrangements, which can complicate processing and "correct" identification. • Broadening and erratic use of arrows, signs, and markings by practitioners makes it difficult for automated systems to properly differentiate between use cases. • Arrow orientation and type matter, as does the layout of the sign legend.
Chris Davies, Potters Industry	**Pavement Markings Guiding Autonomous Vehicles: A Real-World Study** • Davies presented findings of research examining how factors such as reflectivity, contrast, and width affect the performance of machine visions systems. • The range over which vision systems see is adequate for accommodating different speed ranges (30 to 75 mph). • Retroreflectivity is most important at night, while high-contrast ratio is most important during daylight hours. • Wider pavement markings (6 in.) performed as well as or better than 4 in. narrow markings, regardless of stripe color.

Work Zone Management **Jerry Ullman**, *TTI, moderator*	
Speakers	**Summary of Presentation and Question and Answer Session**
Collin Castle, Michigan Department of Transportation	**Michigan DOT Connected Vehicle Work Zone Applications** • Castle provided a summary of Michigan DOT's connected work zone applications. • By 2019, Michigan DOT plans to equip over 350 mi of freeway and arterials to support connected vehicle applications. • One of Michigan DOT's high-priority applications is connected work zone warning and management. • Application use data is entered by field personnel to provide in-vehicle alerts, either through DSRC or mobile devices. • Alerts provide the location of work zones, whether workers are present, and speed advisory information.

Work Zone Management (*continued*) **Jerry Ullman**, *TTI, moderator*	
Speakers	**Summary of Presentation and Question and Answer Session**
Jianming Ma, Texas DOT	**Texas I-35 Connected Work Zone Initiative** • Ma discussed Texas DOT's FRATIS application currently under development for I-35 from Austin to Dallas. • The application involves optimizing freight movement in the corridor by determining the best dispatch time and delivery order for trucks in response to construction activities. • The application uses real-time information about traffic conditions in the corridor, information about planned and ongoing construction activities, and forecasts of lane closure impacts to support optimization. • Phase I concentrates on integrating work zone awareness and impacts on freight planning. • Phase II will expand concepts and information types of other road users in corridor.
Ross Sheckler, iCone Products, Inc.	**What do Automated and Connected Vehicles Need to "Know" About Work Zones?** • Work zones have some unique challenges and issues for CAVs to overcome. • The work zone environment is very fluid—constantly changing as conditions and work activities progress. • Work activities are loosely scheduled by contractors, and subject to changes in construction progress, weather, demands on contractors, and work urgency. • The exact times, locations, durations, and physical activities are not often communicated to the TMC. • Multiple agencies and jurisdictions—public, quasi-public, and private—all conduct work in the roadway environment with no central clearinghouse for disseminating information. • Technologies need to be made affordable for different users, portable in nature, and easy for construction personnel to install and operate.

Digital Mapping and Traveler Information **Carl Anderson**, *FHWA, moderator*	
Speakers	**Summary of Presentation and Question and Answer Session**
Ryoto Shirato, Nisson Motor Company	**Dynamic Map Development Is SIP-adus** • Shirato discussed the need for a hierarchical structure of digital maps layered by time frame. • Digital maps consist of static and dynamic data elements. • The role of agencies is to provide common databases and structured dynamic and static data elements. Service providers can then use common platforms that can be updated to provide value added services. • Structure of dynamic maps includes: – Dynamic point data (vehicles, pedestrians, traffic signals), and line (traffic jam, traffic control) data. – Location reference layer. – Static map data (digital roadway, lane-level). • Dynamic data elements need to be expressed as point or by line.
Ahmed Nasr, HERE	**SENSORIS: Closing the Sensor Data Feedback Loop** • Nasr presented information related to the HERE Interface Specification. • Nasr described the technical requirements of data sets for vehicles submitting on-board vehicle sensor data. • Common interface specification allows data generated by any type of vehicle to be pooled, processed, and analyzed. • Specifications outline reporting frequencies for different classification of data during a path event. – Information about the vehicles (vehicle status, vehicle dynamic, expected state, etc.). – Information about the road (sign recognition, lane boundaries, etc.).
Maxime Flament, ERTICO–ITS Europe	**Regulatory Framework for the Provision of Real-Time Traffic Information in Europe** • Flament discussed the EU's efforts to establish a common framework for providing real-time traffic and travel information. • Member states shall set-up national access points offering a single window of access to data provided by road authorities, road operators, and service providers. • Data and corresponding metadata, including information on data quality to be accessible for exchange and re-use by data service providers. • Data structures include static and dynamic elements needed to support AD in Europe.

Digital Mapping and Traveler Information (*continued*) **Carl Anderson**, *FHWA, moderator*	
Speakers	**Summary of Presentation and Question and Answer Session**
T. Russel Shields, Ygomi, LLC	**Probe Data for Automated Driving** • Vehicle control processes that lead to HAD can be helped by a reliable, up-to-date database of vectors that describe the road network. The needed data input is best collected from vehicle sensors ("probe data"). • The largest barrier to vehicle probe data is communications costs. • Some in-vehicle control processes can be helped by reliable knowledge of the road network. • Many vehicle manufacturer AD experts have determined that they cannot use high-definition maps. • Road network data is best made by collecting probe data from many production vehicles. • Buy-in by vehicle manufacturers, standards bodies, regulators, and the public is crucial. • For HAD, continuous updating of the road network database is likely to be necessary for the service life of the vehicle.

MOVING FORWARD

- Participants suggested that continuing the dialogue on physical infrastructure, work zones, and digital infrastructure at the 2017 AVS would be beneficial.
- Participants identified a number of follow-up activities associated with TRB, including hosting a session at the 95th Annual Meeting summarizing the key elements of this breakout session. Topics identified for further discussion at the TRB Traffic Control Device Committee included developing a call for papers for the 96th Annual Meeting and developing research problem statements focusing on a collaborative review of infrastructure standards and a framework for defining road readiness for AVs. Participants suggested expanding the TRB Symposium on Visibility and Traffic Control Devices, which occurs every 2 years, to include some of the infrastructure topics discussed in the breakout session.

APPENDIX U: BREAKOUT SESSION 21

Traffic Flow of Connected–Automated Vehicles

Organizers

- Xiaopeng (Shaw) Li, University of South Florida;
- Haizhong Wang, Oregon State University;
- Sue Ahn, University of Wisconsin-Madison;
- Robert L. Bertini, Cal Poly, San Luis Obispo;
- Mark Brackstone, Transport Simulation Systems;
- Danjue Chen, University of Wisconsin–Madison;
- Samer Hamdar, George Washington University;
- Steve Mattingly, University of Texas, Arlington;
- Menendez Monica, Swiss Federal Institute of Technology;
- Gabor Orosz, University of Michigan, Ann Arbor; and
- Alireza Talebpour, Texas A&M University.

Reporter

- Xiaopeng (Shaw) Li, University of South Florida

SESSION FOCUS

This breakout session provided an opportunity to bring together the cyber-physical communications, vehicle control, and traffic flow communities to better understand the fundamental characteristics of traffic flow with varying levels of automation and to identify research for developing models to assess the safety, environment, and mobility implications of CAVs. The breakout session focused on the innovative modeling of CAV traffic flow, CAV-based traffic control and management methods, cyber-physical communication frameworks, and related technologies translatable to traffic flow modeling.

Participants discussed potential simulation tools/approaches, data needs, data availability and sources (especially those publicly available), key opportunities for collaboration across research communities, and major challenges in each area. The PowerPoint presentations of most of the presentations are available on the AUVSI website at http://www.automatedvehicles symposium.org/program/breakouts/home.

SESSION SUMMARY

This session included two keynote presentations, three shorter presentations, and a general discussion with all participants.

Autonomous Vehicles and Traffic Physics: Framing the Questions

Hani Mahmassani, *Northwestern University*

Freeways

- Research questions:
  - Implications of connected and autonomous on modeling vehicle–driver behavior and traffic interactions.
  - Automation versus connectivity in terms of safety, throughput (capacity), stability, flow breakdown (reliability), sustainability.
  - Are fundamental diagrams still applicable?

Arterials and Urban Road Junctions

- Will the intersection control systems be as smart as the vehicles and communication systems?
- Will they work under high-volume conditions?

Networks

- Impacts of CVs and AVs on network fundamental diagrams and other network-level relation.
- How are AV fleets routed?

Connected Vehicles Can Increase Throughput and Decrease Delay on Urban Roads

Pravin Varaiya, *University of California, Berkeley* (joint work with J.Lioris, F. Yildiz, R. Pedarsani, D. Farias, A. Kurzhanski, A. Askari)

- ACC can halve saturation headway.
- Network effect: variable length.
- Penetration rate required to eliminate saturation link.
- ACC alone without any changes in infrastructure can increase intersection capacity.
- CACC provides further improvement, but requires additional infrastructure.

SESSION PRESENTATION 1

Research in Connected Automation Applications to Improve Mobility

Osman Altan, *FHWA*

- U.S. DOT is still focusing primarily on Level 1 and Level 2.
- Draft roadmap.
- Early 2020s CV–AV deployed.

- Applications.
- CACC simulation study.
- Can communication and automated control improve lane change and merge maneuvers?
- Looking forward for speed harmonization.
- Eco-approach and departure at a signalized intersection.
- Truck platooning.
- Caltrans/University of California, Berkeley/Volvo (3 truck platooning).

SESSION PRESENTATION 2

Knowledge Gaps in Relation to the Impacts of Automated Driving in Mixed Traffic

Simeon Calvert, *TNO*

- Where we are now: Longitudinal is good; vehicle interaction not so good; SAE>Level 2 we are guessing most of time because no groundtruths?
- Illustration of challenges.
- Anticipatory capabilities.
- Situation and behavior recognition.
- Flexibility of safety protocols.
- Considering other vehicles.
- Equality in relation to conventional vehicles.
- Focus on what we can model.

SESSION PRESENTATION 3

CACC–V2X Solutions to ACC Challenges

Jan-Niklas Meier, *CAMP*

- CACC small-scale test: Multiple makes and models; perform hazard analysis.
- What does a human driver see? They look multiple vehicles ahead when possible (leads to V2V being important).
- Latency (perception reaction time) will still exist in CACC—try to optimize detection, but reaction is likely fixed.
- Faster and more appropriate—reduce system latencies.
- Additional challenges: hazardous situations; different vehicle types; mixed traffic with unequipped vehicles.
- ACC control diagram: radar; CACC; acceleration prediction
- Algorithm development:
  - Foundation layer:
    - Sensor fusion.
    - Lane classification.

▪ Longitudinal control.
– Adaptive time gap.
– Multivehicle look ahead.

SUMMARY OF PANEL DISCUSSION

- Use ACC to reduce reaction time at intersections.
- Methodologies:
  – Measures: safety, throughput (capacity), stability (local and global), flow breakdown (reliability), sustainability.
  – Traffic worthiness to verify that the vehicle will function reasonably. Simulation could be helpful. Verify that it can handle different traffic conditions. Need to define these conditions for traffic worthiness. Use a virtual testbed.
  – Need more field experiments and ground truth data.
- Technological areas:
  – CACC most important to push forward and impact quickly.
  – Lane changing–merging.
  – Communication must be included in analysis and recognize the potential errors (cannot use specific location for vehicles).
- Role of traffic flow modeling:
  – Flexibility, in general and in vehicle/driver behavior reactions latency, in particular.
  – Improved vehicle interactions fundamental issue (human behavior and future vehicle interactions).

Future Research Topics

- GPS positioning resolution for vehicle location and modeling vehicle reactions. Consider GPS error impacts in vehicle interactions.
- CACC platoon breaking by regular vehicles.
- Consider CV–AV as well as CV, AV, and regular vehicles.
- Impact of heterogeneous (safety thresholds) for ACC and CACC.
- Impact of infrastructure and weather on AV controls including these into flow models.
- Without V2V, speeds and system performance will likely have degraded performance?
- Could electric vehicles improve CACC performance because vehicle performance becomes more homogeneous?
- DSRC channels may need to be fully utilized.
- Understand and consider vehicle IDs and platoon IDs.

Key Comments

- Low-hanging fruit available for using ACC to reduce reaction time at intersections
- Methodologies:

  - Simulation–virtual testbed.
  - Traffic worthiness as a criterion to design and verify CAV algorithms.
  - Measures: safety, throughput (capacity), stability (local and global), flow breakdown (reliability), sustainability.
- Implications to traffic flow modeling:
  - Need flexibility in incorporating different communication latencies.
  - Fundamental issues on human driver behavior and vehicle interactions.

MOVING FORWARD

- Need more field experiments and ground truth data to validate model assumptions.
- Cheap and accurate positioning technology is critical to CACC implementation but is yet a challenge.
- DSRC channels may need to be fully utilized.
- Future research:
  - CACC platoon broken by regular vehicles.
  - Various combinations of CAV, CV, AV, and regular vehicles.
  - Impact of heterogeneous vehicles (e.g., trucks and passenger cars).
  - Impact of infrastructure and weather on traffic flow models.
  - Impact of EVs.

APPENDIX V: BREAKOUT SESSION 22

Transformational Technologies

Can Our Research Processes Keep Up?

Organizers

- Mark Norman, TRB;
- Myra Blanco, VTTI;
- Tammy Trimble, VTTI; and
- Jason Bittner, ARA.

Reporters

- Tammy Trimble, VTTI; and
- Mark Norman, TRB.

SESSION FOCUS

A rapidly advancing transportation technology tsunami presents huge challenges and tremendous opportunities for the public sector. If transportation agencies are to facilitate innovation, while at the same time protect public interest, many questions need to be answered. A dynamic, robust, and timely research program can help provide these answers.

Under our traditional research processes, it typically takes years from the inception of a research problem statement to the issuance of the final research report. There are many good reasons for this. However, we need to ask ourselves if this traditional model can continue to meet our needs in this age of rapid change. If not, what can be done about it?

This session built upon prior discussions and efforts underway to address changes to these processes. The session engaged attendees in a discussion of what the next steps need to be to advance transformational technologies in our current processes and identify new activities to accelerate research implementation.

The goal of this session was to identify steps that agencies, TRB, and others can take to help enhance the abilities of transportation agencies to conduct and deliver quality research within timeframes consistent with the pace of rapidly evolving transformational technologies, especially AV technology. Participants addressed the following questions:

- What are the roadblocks faced by public agencies in conducting research and delivering research results?
- Are there options for more rapid response research being employed in the private sector, academia, and other sectors that public agencies might implement to overcome these roadblocks?
- What measures can be taken to help ensure the quality of the research results when employing accelerated research processes?

SESSION SUMMARY

Opening Remarks

Mark Norman, *TRB*

- Overview of where we are and what we have learned to date.
  - Continuing presentations–roundtable discussions.
    - 95th TRB Annual Meeting.
    - AASHTO Standing Committee on Research and Research Administrators Committee Annual Meetings.
    - Automated Vehicle Symposium.
  - Building and sharing list of options for consideration.
- Focus on transformational technologies impacting AVs.
- Need for dialogue.
  - Private-sector investing billions in research and development.
  - Research needed to inform public sector based on fact.
  - Conventional public agency approaches need to be examined.
  - Timeframe may not be compatible.
- Series of questions to consider.
  - Question 1: What are the areas of greatest potential?
    - Research as a continuum rather than discrete individual projects.
    - Accomplish tasks in parallel rather than in series and bring together at end.
    - Use of scenario planning—consider emerging needs rather than simply being reflective and drawing conclusions from past.

Myra Blanco, *VTTI*

- Series of questions to consider.
  - Question 2: What are some ways we can accomplish transformation?
    - Identify and remove roadblocks.
    - Create impetus for change.
    - Practical implications.
    - Leveraging demonstrations (take advantage of scheduled field tests and demonstrations).
  - Question 3: Are there models from other sections that we should be examining?
    - What areas of the research lifecycle provide opportunities for change–improvement?
    - Are there potential models and sources of models (public, private, etc.)?

Jason Bittner, *ARA*

- Participants discussed possible roadblocks.
- Data accessibility (intellectual property concerns) i.e., the ability to share data and have it available.

- Seek ways to speed up processes by which you acquire and analyze data without compromising quality.
- Cooperative agreements.

- Timeline constraints.
- Ability to recruit talent.
- Need simple method for researchers to learn about opportunities and to bid on them without having full-time contracts person.

Participants discussed the following potential solutions.

- Seek ways to speed-up the timeline.
  - Annual deadlines on front end can slow funding – get away from annual discrete deadlines, go more toward National Science Foundation and open call.
  - Perhaps need to examine how other industries and agencies are dealing with these issues; review research to see how much time each step in the life cycle takes to determine where to target.
    - National Academies: 40% of researchers time is spent on meeting administrative requirements.
    - Would agencies know? Perhaps examine budget–timeline (but would only include contracting; publication and miss rest of life cycle).
- Recognize various expectations—government versus private.
  - May need to change mindset—get away from the expectation of a slow timeframe.
  - Government is more bureaucratic to break through. Private funding has a different set of concerns.
  - Private near-term needs more unstable than longer-term where you can guide company toward that. Sometime more deliberative thinking is necessary.
- Recognize different strata of research projects—not one model fits all.
  - Need to have oversite to ensure public money spent wisely (especially funding with little national visibility or accountability).
  - Need clearer definitions of research strata – some larger, complex projects and other smaller, shorter timeframe projects.
    - Large projects not amenable to 1 year/quicker turnaround. Rather, break into smaller segments to enable sharing of results in a timelier manner.
    - Or not smaller chunks, perhaps—larger project that is not so constrained—not telling researchers how to achieve ultimate outcome and breaking into chunks. Allow for tools and questions to keep evolving.
    - Research program to address transformational technologies—shorten the time up front—depending on public or private entity.
- Recognize need for a different research culture that is more nimble—reflective of transformational technologies.
  - Complete pieces of work to provide feedback on interim/early reporting so that if something changes, can adapt.
  - Slow research program implementation to deal with infrastructure – which is longer—something new for transformational technologies.
- Make research needs process more continuous and more visible.

– Social media-based pulse of what research needs are—these are the top 10 issues this month: research needs dashboard.

Participants discussed the following different research models.

- EU commission is working on work problem for 2018, 2019, and 2020.
  - Ask all for ideas.
  - Do higher risk or greater good research; do not get into area where the private sector is considering (e.g., sensors).
- Government/industry/university partnerships: Enterprise projects distributed geographically. Draw best of UTC–DOT funding – has transparency.
- NHTSA—green—like the alcohol project.
- New contracting mechanisms reflective more of private sector.
  - Need mechanism to allow for stoppage—regroup/reassess—without needing stop work orders, new statement of work.
  - Would need to be balanced with need to prevent abuse.

Participants identified the following ways to improve the research process.

- Seek ways to speed up the timeline.
- Recognize various expectations: government versus private.
- Need clearer definitions of strata of research—some larger, complex projects and other smaller, shorter timeframe projects.
- Recognize need for a different research culture that is more nimble—reflective of transformational technologies.
- Seek ways to make the research needs process more continuous and more visible, e.g., a research needs dashboard.

MOVING FORWARD

- Comments and feedback will be synthesized.
- Synthesis will be presented to the TRB Conduct of Research Committee.
- Will seek additional ways to share results: summary flyer, more in-depth report, distribute/publicize, present at webinars, sessions, etc.

APPENDIX W

List of Acronyms

AAA	American Automobile Association
AASHTO	American Association of State Highway and Transportation Officials
AAVMA	American Association of Motor Vehicle Administrators
ABM	Activity-Based Model
ACC	Adaptive Cruise Control; Automatic Cruise Control; Active Cruise Control
ACRP	Airport Cooperative Research Program
AD	Automated Driving
AdaptIVe	Automated Driving Applications and Technologies for Intelligent Vehicles
ADAS	Advanced Driver Assistance Systems
AOD	Autonomous-Oriented Development
ARC	Atlanta Regional Commission
ART	Advanced Rapid Transit
ARTS	Automated Road Transport Systems
ASCC	Advanced Smart Cruise Control
ASD	Aftermarket Safety Devices
ATCMTD	Advanced Transportation and Congestion Management Technology Deployment
AUVSI	Association for Unmanned Vehicle Systems International
AV	Automated Vehicle
AVCS	Advanced Vehicle Control Systems
AVS	Automated Vehicle Symposium
BART	Bay Area Rapid Transit
BRT	Bus Rapid Transit
CACC	Cooperative Adaptive Cruise Control
C-AD	Connected-Automated Driving
Caltrans	California Department of Transportation
CAVs	Connected and Automated Vehicles
CCTA	Contra Costa Transportation Authority
CEO	Chief Executive Officer
C-ITS	Cooperative Intelligent Transport System
CO	Carbon Monoxide
CO_2	Carbon Dioxide
COMPANION	Cooperative Dynamic Formation of Platoons for Safe and Energy-Optimized Goods Movement
ConOps	Concept of Operations
CoVASS	Cooperative Vehicle Adaptive Control System
CPU	Central Processing Units
CV	Connected Vehicle
DOE	Department of Energy

DMV	Department of Motor Vehicles
DRCC	Dynamic Radar Cruise Control
DSRC	Dedicated Short Range Communication
DVI	Driver Vehicle Interface Page
EC	European Commission
EU	European Union
FAA	Federal Aviation Administration
FDOT	Florida Department of Transportation
FHWA	Federal Highway Administration
FMVSS	Federal Motor Vehicle Safety Standards
FTA	Federal Transit Administration
GCDC	Grand Cooperative Driving Challenge
GHG	Greenhouse Gas
GPS	Global Positioning System
HAD	Highly Automated Driving
HAV	Highly Automated Vehicle
HDV	Heavy-Duty Vehicles
HMI	Human Machine Interface
HOV	High-Occupancy Vehicle
ICC	Intelligent Cruise Control
ICT	Information and Communication Technologies
IRB	Institutional Review Board
ITS	Intelligent Transportation System
ITS JPO	Intelligent Transportation Systems Joint Program Office
LADOT	Los Angeles Department of Transportation (California)
LiDAR	Light Detection and Ranging
LRT	Light Rail Transit
MMITSS	Multimodal Intelligent Traffic Signal System
MOD	Mobility on Demand
MoLIT	Ministry of Land, Infrastructure, and Transport (Japan)
MPO	Metropolitan Planning Organization
NACTO	National Association of City Transportation Officials
NASA	National Aeronautics and Space Administration
NCHRP	National Cooperative Highway Research Program
NDOT	Nevada Department of Transportation
NextGen	Next Generation Air Traffic Control System
NHTSA	National Highway Traffic Safety Administration
NOC	Network Operations Center
NSF	National Science Foundation
NTC	National Transport Commission (Australia)
NVTA	Northern Virginia Transportation Authority
OBD	On-Board Diagnostic
OECD	Organization for Economic Cooperation and Development
OEMs	Original Equipment Manufacturers
OSU	The Ohio State University
PED-SIG	Mobile Accessible Pedestrian Signal System

PEGASUS	Project for establishing generally accepted quality criteria, tools, and methods, as well as scenarios and situations
PHEV	Hybrid Plug-In Electric Vehicles
PNDs	Personal Navigation Devices
PPU	Pay-Per-Use
PROSPECT	PROactive Safety for Pedestrians and Cyclists
PRT	Personal Rapid Transit
R&I	Research and Implementation
RAC	Research Administrators Committee
SAE	Society of Automotive Engineers
SAEV	Shared Autonomous Electric Vehicles
SAFE	Saving America's Future Energy
SCOR	Standing Committee on Research
SDV	Self-Driving Vehicles
SIP-adus	Strategic Innovation Promotion Program Innovation of Automated Driving for Universal Services
SIRIA	Strategic Transport Research and Innovation Agenda
SMART	Systems and Modeling for Accelerated Research in Transportation
SMS	Safety Management System
SP3	Sub Project 3
SPARC	Sports Pavilion and Automotive Research Complex
TAAS	Transportation-as-a-System
TCRP	Transit Cooperative Research Program
TfL	Transport for London
3D	Three-Dimensional
TMC	Traffic Management Center
TNC	Transportation Network Company
TOD	Transit-Oriented Development
TRB	Transportation Research Board
TRC	Transportation Research Center
TSPS	Traffic Signal Prediction Systems
TTI	Texas A&M Transportation Institute
U.S.	United States
UAVs	Unmanned Aerial Vehicles
UDDS	Urban Dynamometer Driving Schedule
UK	United Kingdom
U.S. DOT	United States Department of Transportation
V2I	Vehicle-to-Infrastructure
V2V	Vehicle-to-Vehicle
V2P	Vehicle-to-Pedestrian
V2X	Vehicle-to-Everything
V&V	Verification and Validation
VMT	Vehicle Miles Traveled
VRU	Vulnerable Road Users
VTTI	Virginia Tech Transportation Institute
WASP	Wallenberg Autonomous Systems Program

The National Academies of

SCIENCES • ENGINEERING • MEDICINE

The **National Academy of Sciences** was established in 1863 by an Act of Congress, signed by President Lincoln, as a private, nongovernmental institution to advise the nation on issues related to science and technology. Members are elected by their peers for outstanding contributions to research. Dr. Marcia McNutt is president.

The **National Academy of Engineering** was established in 1964 under the charter of the National Academy of Sciences to bring the practices of engineering to advising the nation. Members are elected by their peers for extraordinary contributions to engineering. Dr. C. D. Mote, Jr., is president.

The **National Academy of Medicine** (formerly the Institute of Medicine) was established in 1970 under the charter of the National Academy of Sciences to advise the nation on medical and health issues. Members are elected by their peers for distinguished contributions to medicine and health. Dr. Victor J. Dzau is president.

The three Academies work together as the National Academies of Sciences, Engineering, and Medicine to provide independent, objective analysis and advice to the nation and conduct other activities to solve complex problems and inform public policy decisions. The Academies also encourage education and research, recognize outstanding contributions to knowledge, and increase public understanding in matters of science, engineering, and medicine.

Learn more about the National Academies of Sciences, Engineering, and Medicine at **www.national-academies.org.**

The **Transportation Research Board** is one of seven major programs of the National Academies of Sciences, Engineering, and Medicine. The mission of the Transportation Research Board is to increase the benefits that transportation contributes to society by providing leadership in transportation innovation and progress through research and information exchange, conducted within a setting that is objective, interdisciplinary, and multimodal. The Board's varied committees, task forces, and panels annually engage about 7,000 engineers, scientists, and other transportation researchers and practitioners from the public and private sectors and academia, all of whom contribute their expertise in the public interest. The program is supported by state transportation departments, federal agencies including the component administrations of the U.S. Department of Transportation, and other organizations and individuals interested in the development of transportation.

Learn more about the Transportation Research Board at **www.TRB.org**.